Biotechnology
Theory and Techniques

Volume II

Biotechnology
Theory and Techniques
Volume II

Genetic Engineering
Mutagenesis
Separation Technology

This book was written with support, in part,
from the National Science Foundation

Editor

Jack G. Chirikjian
Georgetown University School of Medicine, Washington, D. C.
Chairman, EDVOTEK, Inc., Rockville, Maryland

Associate Editors

Edward C. Kisailus
Canisius College
Buffalo, New York

Baldwin King
Drew University
Madison, New Jersey

Robert Krasner
Providence College
Providence, Rhode Island

Harley Mortensen
SW Missouri State University
Springfield, Missouri

Managing Editor

Karen M. Graf
EDVOTEK, Inc.

CBS

CBS PUBLISHERS & DISTRIBUTORS PVT. LTD.
NEW DELHI • BENGALURU • CHENNAI • KOCHI • MUMBAI • PUNE

CBS Publishers ISBN : 978-81-239-1630-9 (PB)
ISBN : 978-81-239-1717-7 (HB)
Jones & Bartlett ISBN : 0-86720-896-1

First Indian Reprint : 2009

This edition has been reprinted in India by special arrangement with Jones & Bartlett Publishers Inc., USA

Published by Satish Kumar Jain and produced by V.K. Jain for
CBS Publishers & Distributors Pvt. Ltd.,
CBS Plaza, 4819/XI Prahlad Street, 24 Ansari Road, Daryaganj,
New Delhi - 110002, India. • Website: www.cbspd.com
e-mail: delhi@cbspd.com, cbspubs@airtelmail.in
Ph.: 23289259, 23266861, 23266867 • Fax: 011-23243014

Branches:

- ***Bengaluru:*** Seema House, 2975, 17th Cross, K.R. Road, Bansankari 2nd Stage, Bengaluru - 560070
  • Ph.: +91-80-26771678/79 • Fax: +91-80-26771680
  • E-mail: cbsbng@gmail.com, bangalore@cbspd.com
- ***Pune:*** Bhuruk Prestige, Sr. No. 52/12/2+1+3/2, Narhe, Haveli (Near Katraj-Dehu Road By-pass), Pune - 411041
  • Ph.: +91-20-64704058/59, 020-32392277 • E-mail: pune@cbspd.com
- ***Kochi:*** 36/14, Kalluvilakam, Lissie Hospital Road, Kochi - 682018, Kerala • Ph.: +91-484-4059061-65
  • Fax: +91-484-4059065 • E-mail: cochin@cbspd.com
- ***Chennai:*** 20, West Park Road, Shenoy Nagar, Chennai - 600030
  Ph.: +91-44-26260666, 26208620 • Fax: +91-44-42032115
  • E-mail: chennai@cbspd.com
- ***Mumbai:*** 83-C, Ist Floor, Dr. E. Moses Road, Worli, Mumbai-400 018, Maharashtra Ph.: +91-9833017933, 022-24902340/24902341
  • E-mail: mumbai@cbspd.com

Printed at :
India Binding House, Noida (UP)

CONTRIBUTORS

Jack G. Chirikjian
Georgetown University School of Medicine, Washington, D. C.
Chairman, EDVOTEK, Inc., Rockville, Maryland

Dennis Bogyo
Valdosta State College
Valdosta, Georgia

John Boyle
Cerritos College
Norwalk, California

Audrey Brown
Seattle Central Community College
Seattle, Washington

Jordan Choper
Montgomery College
Takoma Park, Maryland

G. Bruce Collier
Trevigen, Inc.
Gaithersburg, Maryland

Richard Echols
Southern University
Baton Rouge, Louisiana

Mark A. Holland
Salisbury State University
Salisbury, Maryland

Barbara Jones
Southwestern Adventist College
Keene, Texas

Baldwin King
Drew University
Madison, New Jersey

Edward Kisailus
Canisius College
Buffalo, New York

Karen K. Klyczek
University of Wisconsin-River Falls
River Falls, Wisconsin

Robert Krasner
Providence College
Providence, Rhode Island

Malethu T. Mathew
Virginia Union University
Richmond, Virginia

Harley Mortensen
SW Missouri State University
Springfield, Missouri

Ah-Kau Ng
University of Southern Maine
Portland, Maine

Mark Petersen
Blue Mountain Community College
Pendleton, Oregon

E. Robert Powell
Central Oregon Community College
Bend, Oregon

Geraldine Ross
Highline Community College
Des Moines, Washington

Anthony Sena
Northern New Mexico Community College
Espanola, New Mexico

Ellie Skokan
Wichita State University
Wichita, Kansas

Kathy Steinert
Bellevue Community College
Bellevue, Washington

Peter Woodruff
Champlain Regional College
St. Lambert, Quebec

CONTENTS

TO THE INSTRUCTOR

The work presented in the two volumes of Biotechnology Theory and Techniques is the collaborative effort of over twenty undergraduate science faculty. The undergraduate faculty represent various two-year and four-year colleges in several regions of the United States and Canada. The common goal of the faculty contributors was to develop a publication containing unique and flexible laboratory activities that focus on the theory and practice of biotechnology for students.

The experiments presented in both volumes are designed to utilize many resources readily available in most science departments. Care was taken to avoid duplicating research protocols and laboratory experiments that are available in a variety of existing laboratory manuals. We believe that Biotechnology Theory and Techniques is a unique and practical compilation of experiment modules that are safe, accurate, and appropriate for implementation in academic science laboratories.

The topics and experiments included are deemed by the faculty to represent major areas of biotechnology, but by no means address all aspects of the continuously evolving technology. The two volumes are not designed for any specific science course. The strategy is to allow the flexibility for selecting various modules to be implemented in life science courses, and wherever else they are applicable. Alternate approaches are included.

As Editor of the two volumes of Biotechnology Theory and Techniques, my role is primarily that of a catalyst. I am fortunate to be able to assemble a creative and dedicated group of science college faculty whose goal is to enhance a classroom experience for student interested in learning about the science of biotechnology.

TO THE STUDENT

Although the protocols, techniques, and procedures organized in this manual have been developed over years of work by scientific investigators, these laboratories should not be considered as canned, nor will they succeed in every way, every time. They should be looked upon as experiences leading to adventure, challenge, and discovery.

Your instructor may want you to keep a separate duplicate laboratory notebook for your write-ups. The write-up is a very important element in the actual learning of laboratory science. It is precisely in the doing of laboratories such as those contained herein that you learn how to do research and develop a sense of how to approach a problem, seeking and recording data as you proceed. You should not be discouraged if you do not get exact results in each experiment. Always ask yourself: What could have been the reason for the different results? What could I do differently that might lead to the desired outcome?

Always remember that it has taken years of invested time for teachers and researchers to develop the laboratory experiences that you will be attempting with this manual. These exercises are not chosen for ritualistic value, but rather to expose you to the wonderment of the experimental spirit that is the essence of science. Enjoy the adventure.

ACKNOWLEDGMENTS

Thank you to Jones and Bartlett, our publisher, for ensuring the earliest possible publication of our work. We appreciate the marketing wisdom of our sponsoring editor, David Phanco and the efforts of editorial assistant, Deborah Haffner as well as the production editors, Mary Cervantes-Sanger and Nadine Fitzwilliam.

I am also grateful to members of the group for their individual and collective contributions. A special thanks is extended to the staff at EDVOTEK who contributed to this work, and to Edward Kisailus, Baldwin King, Robert Krasner, and Harlye Mortensen, the four associate editors of the project. Special recognition goes to Karen Graf, managing editor, whose energy and dedication has made the completion of the project possible. Recognition is also extended to Kathy Gilbert, editorial assistant, for her contribution to the presentation of the experiment modules. A final special note of appreciation goes to the National Science Foundation, which provided support for initiating the project.

Jack G. Chirikjian, Ph.D.
Professor of Biotechnology and Molecular Biology
Georgetown University School of Medicine, Washington D.C.
Chairman, EDVOTEK, Inc. Rockville, Maryland

UNIT 1

DNA Isolation

UNIT 1: DNA ISOLATION

MODULE 1: DNA ANALYSIS BY GEL ELECTROPHORESIS

Kathy Steinert
Geraldine Ross
John S. Boyle

* Introduction
* Safety Guidelines
* Experimental Outline
* Materials
* Method
* Results

Introduction

Agarose gel electrophoresis is a widely used method that separates molecules based upon charge, size and shape. It is particularly useful in separating charged biomolecules such as DNA, RNA and proteins.

Agarose gel electrophoresis possesses great resolving power, yet is relatively simple and straightforward to perform. The gel is made by dissolving agarose powder in boiling buffer solution. The solution is then cooled to approximately 50°C and poured into a mold where it solidifies. The gel is submerged in a buffer-filled chamber which contains electrodes.

DNA samples are prepared for electrophoresis by mixing them with solutions containing glycerol or sucrose. This makes the samples denser than the electrophoresis buffer. These samples can then be loaded with a micropipet or transfer pipet into wells that were created in the gel by a template during casting. The dense samples sink through the buffer and remain in the wells.

A direct current power supply is connected to the electrophoresis apparatus and current is applied. Charged molecules in the sample enter the gel through the walls of the wells. Molecules having a net negative charge migrate towards the positive electrode (anode) while net positively charged molecules migrate towards the negative electrode (cathode). Within a range, the higher the applied voltage, the faster the samples migrate. The buffer serves as a conductor of electricity and to control the pH. The pH is important to the charge and stability of biological molecules.

Agarose is a polysaccharide derivative of agar. The agarose gel contains microscopic pores which act as a molecular sieve. The sieving properties of the gel influences the rate at which a molecule migrates. Smaller molecules move through the pores more easily than larger ones. Molecules can have the same molecular weight and charge but different shapes. Molecules having a more compact shape (a sphere is more compact than a rod) can move more easily through the pores.

Factors, such as charge, size and shape, together with buffer conditions, gel concentrations and voltage, affects the mobility of molecules in gels. Given two molecules of the same molecular weight and shape, the one with the greater amount of charge will migrate faster. In addition, different molecules can interact with agarose to varying degrees. Molecules that bind more strongly to the agarose will migrate more slowly.

The objective of this module is to familiarize students with the principles of agarose gel electrophoresis by preparing high molecular weight DNA for electrophoresis. Although chromosomal DNA

Introduction, continued

will be subjected to electrophoretic separation, plasmid DNAs and smaller DNA digests are also effectively separated by this procedure. DNA that has been extracted by precipitation from calf thymus, pea plants or *E. coli,* or by spooling from EDVOTEK experiment #107, may be re-dissolved and loaded onto an agarose gel and electrophoresed. Following the electrophoresis, the DNA may be visualized by staining the DNA with Methylene Blue Plus™ or ethidium bromide staining reagents.

Chromosomal DNA that you have previously extracted is of very high molecular weight. There may have been some shearing, but the bulk of your solution will consist of very large molecules. Since the migration rate of DNA through agarose during electrophoresis is inversely proportional to its molecular weight, your sample should migrate a very short distance. (Note: If you are electophoresing chromosomal DNA along with *Hind* III standard markers, you will not be able to estimate the size of your chromosomal DNA by placement of its migration upon the standard curve. This is because larger DNA fragments, in excess of 20 kilobases in length, have different charge:mass ratios and tend to migrate faster than one would predict. This effect is potentiated by increasing gel porosity and electric field strength.)

In this laboratory module you will prepare an agarose gel of 0.8%. You will then load your chromosomal DNA into one of the wells and electrophorese it for at least 1 1/2 hours. Following electrophoresis you will stain the gel in methylene blue stain or ethidium bromide enabling visualization of the DNA in the gel.

Safety Guidelines

Boiling agarose can spatter and cause severe burns. When heating agarose wear safety goggles and hot gloves. Latex gloves should be worn throughout the procedure, especially when handling solutions or gels containing methylene blue. Prior to turning electrophoresis power pack on, be sure that the chamber is level and that the work surface is dry. Wear gloves and safety goggles when operating electrophoresis apparatus.

Ethidium bromide is a powerful mutagen. Wear eye and skin protection (gloves) while working with ethidium bromide and UV radiation emitted from the transilluminator. Ethidium bromide solution should be treated as hazardous waste and disposed of properly.

Experimental Outline

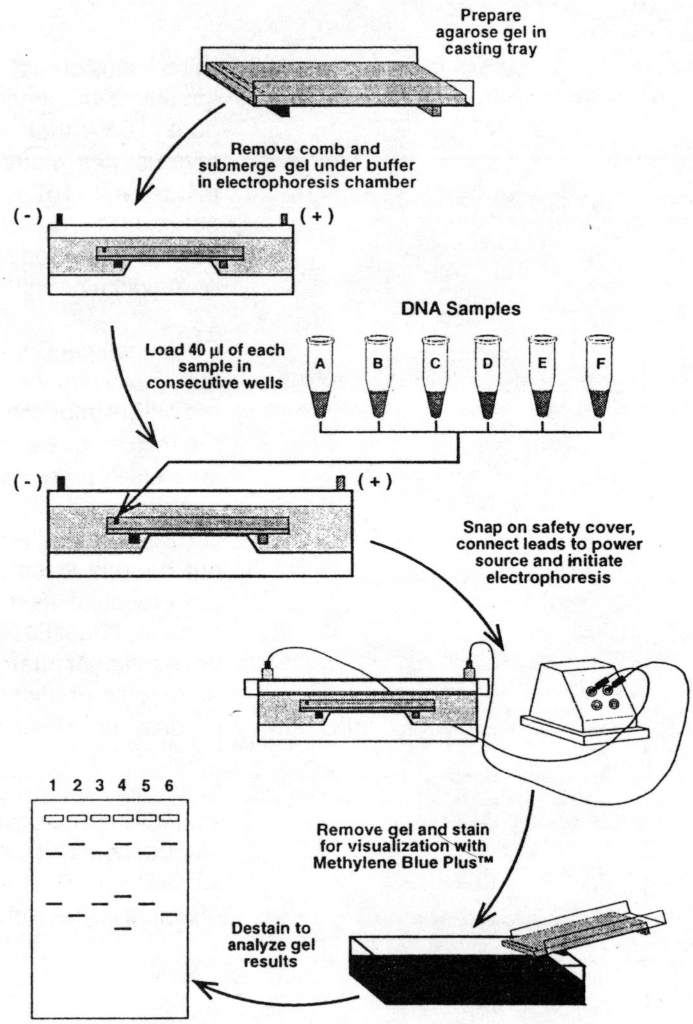

Materials

Micropipet with disposable tips
Microfuge test tubes (1.5 ml)
Electrophoresis chamber
Electrophoresis power pack
Hot plate or microwave oven
Safety goggles
Insulated gloves
Laboratory tape
100 ml beakers
Balance weighing paper
Spatula
Staining tray

Materials, continued

Agarose (electrophoresis grade)
Electrophoresis Buffer (1x working buffer)
    (EDVOTEK 50x TAE buffer or 10x TBE buffer)
Tracking Dye/Loading Buffer
    40% (w/v) sucrose or glycerol
    0.25% bromophenol blue
    TAE or TBE (1x)
Methylene Blue Plus™ Stain (EDVOTEK, Inc.)
    or Ethidium Bromide (10 mg/ml)
Distilled water
100 ml graduated cylinders
Saran Wrap
Latex gloves

Methylene Blue Plus™ Stain:

Methylene Blue Plus™ is a specially formulated DNA staining reagent which may be purchased from EDVOTEK, Inc. P.O. Box 1232 West Bethesda, Maryland 20827-1232 1-800-EDVOTEK (1-800-338-6835)

Tris-borate Buffer (TBE, 10x pH 8.3) for electrophoresis can be made from the following recipe:

0.89 M Tris base (108 gm)
0.89 M Boric Acid (55 gm)
0.02 M $Na_2EDTA \cdot 2H_20$ (9.3 gm)
ddH_20 to 1 liter

Alternatively, Tris-acetate-EDTA may be used for electrophoresis. A 50-fold concentrated TAE buffer can be purchased from EDVOTEK

Pre-lab Preparation

The instructions contained in this module are specific to the utilization of EDVOTEK electrophoresis units. For use of other models, some modifications may be necessary.

1. Agarose Gel Preparation

Your individual schedule and time requirements will determine when the agarose gel is prepared. It takes approximately 30 to 40 minutes to prepare the gel (generally 20 minutes of this time is required for solidification). The solidified gel can be stored in the covered electrophoresis apparatus under buffer for 2 or 3 days before the laboratory. Samples must be electrophoresed immediately after they are loaded since they will diffuse out of the well.

Pre-lab Preparation, continued

2. Placement of Combs

Place one comb in the end notches of the gel bed or platform. There should be a small space between the bottom of the teeth and the bed. Make sure the combs are placed evenly across the bed.

3. Gel Concentration

The agarose gel concentration required for most agarose gel electrophoresis experiments is 0.8% weight by volume. Table A contains guidelines for quantities of agarose, concentrated buffer and distilled water required for preparing the gel. To enhance migration of the high molecular weight chromosomal DNA, gels of lower concentration may be used. However, as gel concentration decreases the fragility of the gel increases.

4. Electrophoresis (chamber) Buffer

EDVOTEK kits contain 50x concentrated electrophoresis buffer. Add 1 volume buffer to every 49 volumes of distilled or deionized water (see Table B).

If 10x TBE buffer is used, mix 50 ml of concentrated TBE buffer with 450 ml distilled water to make 500 ml chamber buffer.

Table A: Guidelines for preparing Individual 0.8 % Agarose Gels

EDVOTEK Model #	Approximate Gel Bed Dimensions (W x L)	Amt of Agarose	Volume of Buffer (50x TAE)	+	Distilled Water	=	Total Volume
M6 or M36	7 x 7 cm	0.24 gm	0.6 ml		29.4 ml		30 ml
M12	7 x 15 cm	0.48 gm	1.2 ml		58.8 ml		60 ml
M20	10.5 x 14 cm	0.8 gm	2.0 ml		98.0 ml		100 ml

Table C: Guidelines for Time and Voltage

Volts	Recommended Time: Minimum	Recommended Time: Optimal
15	8.0 hrs	10.0 hrs
25	2.0 hrs	3.0 hrs
50	1.0 hr	2.0 hrs
75	40 min	1.5 hrs
125	30 min	1.0 hrs

Table B: Electrophoresis (Chamber) Buffer

EDVOTEK Model #	Volume of Buffer (50x TAE)	+	Distilled Water	=	Total Volume
M6	4 ml		196 ml		200 ml
M12	6 ml		294 ml		300 ml
M20	8 ml		392 ml		400 ml
M36	8 ml		392 ml		400 ml*

*Additional buffer may be required if running the Model #M36 with less than 6 gel beds.

Pre-lab Preparation, continued

5. Electrophoresis Time and Voltage

The minimum time and voltage to show migration of chromosomal DNA is 1 hour at 50 volts (see Table C). The power supply may be connected to a simple household light timer set to go off at a designated time. The gel can sit overnight in the apparatus under buffer, then stained the following day.

Method

Casting an Agarose Gel

1. When using a gel tray that is open at two ends, begin by sealing the open ends with 3/4" wide tape. (Some electrophoresis units may have rubber end caps, which can be used in place of tape.) Extend the tape over the sides and bottom edges of the tray. If using a gel tray that has two ends that slide up, slide them down to seal the tray.

2. If using an EDVOTEK electrophoresis apparatus, refer to Table A to determine the amounts of agarose, concentrated TAE buffer and distilled or deionized water to measure into a 250 ml flask. This step may be performed in one of two ways:

   a. Warm the mixture in the flask in a 100°C water bath, until the agarose is dissolved.

   b. If a microwave is available, heat the mixture in the flask for 2 minutes on high, or until the agarose is completely dissolved. Take care not to let the solution boil over.

3. Cool the agarose solution either on a ring stand until its temperature reaches 50°C, or in a 50°C waterbath. Be sure to swirl the flask as the solution cools.

4. To prevent liquid agarose from leaking out of the taped ends of the gel tray, seal the sticky side with a thin line of cooled agarose solution.

5. With gel tray on a level surface, pour the rest of the flask contents into the gel tray. Place the comb into the notched end of the tray and let the solution set.

6. If bubbles form while pouring the gel, use a pipette tip to push them toward the edge of the gel. If this is not done, the current will not run uniformly through the gel and the results may be disappointing.

Method, continued

Casting an Agarose Gel, continued

7. After the gel has solidified, it will become cloudy. Remove the comb and the tape from the tray. Place the tray on the raised platform within the apparatus. (Gels can be refrigerated in this form within a covered apparatus for 2 - 3 days.) Do not let the gel slide off the tray during this transfer.

8. Mix and fill the electrophoresis apparatus with the appropriate amount of chamber buffer (Table B).

9. Be sure that the gel is placed within the apparatus with the sample wells closest to the negative (black) electrode. Otherwise, the sample will migrate into the chamber buffer, instead of going toward the positive (red) electrode.

Preparation and Loading of Sample into Gel

1. To a 1.5 ml microcentrifuge tube, add 5 microliters of 10x gel loading solution and 45 microliters of re-dissolved DNA. Mix by tapping on table top.

2. Load up to 40 microliters of DNA/Gel loading solution mixture to fill a well in the agarose gel. If you are also running DNA fragments with a *Hind* III marker, load the chromosomal DNA into the well to the right of the marker.

Running the Gel

1. Carefully snap the cover all the way down onto the electrode terminals. On EDVOTEK electrophoresis units, negative and positive terminals are indicated on the cover and apparatus chamber.

2. Insert the plug of the black wire into the black input of the power supply (negative input). Insert the plug of the red wire into the red input of the power supply (positive input.)

3. Turn on and set the power supply at 50 volts. The minimum running time for adequate separation at 50 volts is 1 hour with an optimum time of approximately 2-3 hours. At higher voltages, the running time is reduced. The electrophoresis should be terminated before the tracking dye reaches the end of the gel. When the current is flowing, you should see bubbles forming on the electrodes.

Method, continued

Running the Gel, continued

4. After the electrophoresis is completed, turn off the power, unplug the unit, disconnect the leads and remove the cover.

   Note: After the electrophoresis is completed, the gel can be wrapped and refrigerated until you are ready to stain. Do not freeze agarose gels. For optimal results, the gel should be stained right after electrophoresis.

5. Remove the gel from its bed. Keep your hands on each end of the gel to prevent it from slipping off the bed during manipulations. Remove a small slice of gel from the upper right hand corner so you can keep track of right and left.

6. Transfer the gel to the gel staining tray for staining.

Staining With Methylene Blue Pus™

1. Cover gel with methylene blue staining solution and leave for 20-30 minutes.

2. Following staining, remove methylene blue solution from the staining chamber. (It may be reused up to five times.) Cover the gel with distilled water to destain. DNA should become visible within 30-60 minutes. Continue destaining for approximately 2-4 hours. Destaining may be continued overnight. Do not destain the gel in a second change of water since this will cause the bands to fade.

3. After destaining, carefully remove the gel and slide it onto the surface of a white light source for visualization.

4. Methylene blue binds ionically to the DNA and can be seen with the naked eye. Unlike ethidium bromide, it is not hazardous, but does not detect DNA levels as low as can be detected by ethidium bromide.

Ethidium Bromide Staining

ETHIDIUM BROMIDE IS A POWERFUL MUTAGEN AND SHOULD ALWAYS BE HANDLED USING GLOVES.

1. Carefully transfer the gel to an ethidium bromide stain prepared by adding 10 μl of ethidium bromide stock (10 mg/ml) to 100 ml of water.

Method, continued

Ethidium Bromide Staining, continued

2. Stain 10 minutes followed by a 10 minute destain in water.

3. In order to detect DNA using ethidium bromide, a UV source such as a transilluminator is required. Ethidium bromide intercalates between the bases of the DNA and will fluoresce when stimulated by ultraviolet light. Adequate eye and skin protection against UV radiation is required when detecting DNA stained with ethidium bromide. After destaining, place the gel on the transilluminator and view through protective goggles or face shield.

4. Usually, the gel is photographed to obtain a permanent record of the band pattern in the gel. A suitable camera is the Polaroid DS34 Direct Screen Instant Camera using Polaroid 667 film. Alternatively, an image of the fluorescent gel can be traced on transparency film to provide a permanent record of the experiment.

Results

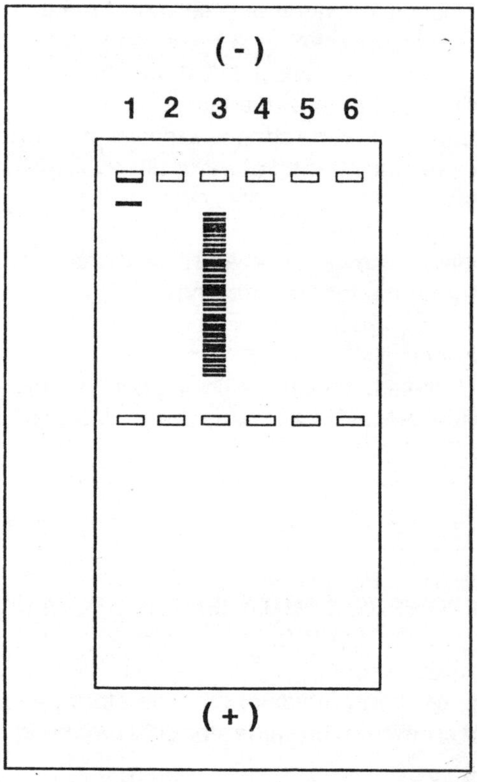

The figure to the left is an idealized schematic showing the relative mobility of chromosomal DNA samples. The idealized schematic shows the relative positions of the bands, but are not depicted to scale.

Lane 1: Intact Chromosomal DNA

In this example, a band near the edge of the well may sometimes be observed. This extra band can be due to residual bound proteins that retard migration of the DNA into the gel.

Lane 3: Example of badly sheared chromosomal DNA

Methylene blue binds ionically to the DNA and can be seen with the naked eye. It is less hazardous, but does not detect DNA levels as low as can be detected by ethidium bromide. An image of the gel can be traced on transparency film to provide a permanent record of the experiment.

Results, continued

Gel Electrophoresis Data Sheet

Name: ______________________________ Partners: ______________________________

Gel Preparation: _____ % gel Agarose powder _____ g

Concentrated buffer _____ ml Sample volume _____ µl

Run conditions: _____ min _____ volts _____ mm Marker migration

Stain: ____

Well Data

Well Number	1	2	3	4	5	6	7
Contents	______	______	______	______	______	______	______
# of bands	______	______	______	______	______	______	______
Migration (mm)	______	______	______	______	______	______	______

Well Number	8	9	10	11	12	13	14
Contents	______	______	______	______	______	______	______
# of bands	______	______	______	______	______	______	______
Migration (mm)	______	______	______	______	______	______	______

Comments:

UNIT 1: DNA ISOLATION

MODULE 2: PLASMID ISOLATION

Harley Mortensen

- Introduction
- Safety Guidelines
- Experimental Outline
- Materials
- Method
- Results

Introduction

The objectives of this module are to isolate plasmid DNA from a bacterial host cell and to learn basic techniques and procedures used in recombinant DNA work. Plasmids are extrachromosomal, circular DNA capable of autonomous replication. They frequently carry genes for antibiotic resistance (which serve conveniently as screening markers) and can be used to transform host cells. Many plasmid derivatives have been developed which can be used as recipients for inserting DNA segments from other sources into specific regions of the plasmid. Thus, plasmids are widely used as vectors.

Isolation of plasmid DNA is one of the first procedures presented to students learning the basic techniques of recombinant DNA work. In contrast to chromosomal DNA and large plasmids, the types of plasmids used here are sturdy (not subject to breaking by shearing action during isolation) and easily reconstituted into the native conformation after treatments which leave the other forms of DNA denatured.

Two procedures are presented:

1. A rapid isolation developed for small amounts (miniprep) of DNA which is not very pure but still suitable for some purposes;

2. A scaled up protocol for larger amounts of DNA of higher purity. This procedure calls for 25-40 ml of an overnight culture. As proficiency develops, the culture volumes may be reduced to as low as 1 ml with proportionate reductions in reagents.

Safety Guidelines

Use gloves and goggles when handling the phenol:chloroform mixture. Phenol causes burns when in contact with tissue. Avoid breathing or contact with chloroform. Phenol and chloroform should be treated as hazardous waste and disposed of properly. Although there are procedures that avoid phenol to extract plasmid DNA, this reagent is still widely used in biotechnology laboratories.

After gel electrophoresis of the plasmid, staining of DNA is required for visualization. DNA may be stained with either EDVOTEK's Methylene Blue Plus™, or with ethidium bromide for visualization under UV fluorescent light. Ethidium bromide is a powerful mutagen. Wear eye and skin protection (UV protective goggles and gloves) while working with ethidium bromide and UV radiation emitted from the transilluminator. Ethidium bromide solution should be treated as hazardous waste and disposed of properly.

Experimental Outline

Plasmid Miniprep:

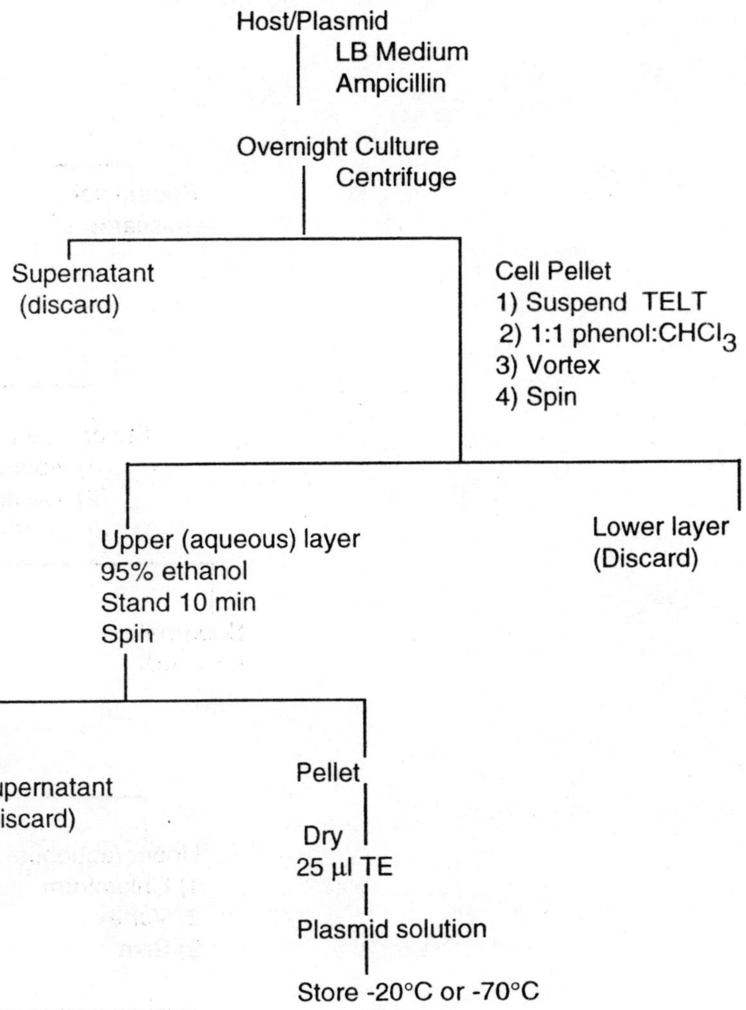

Experimental Outline, continued

Large Scale Prep

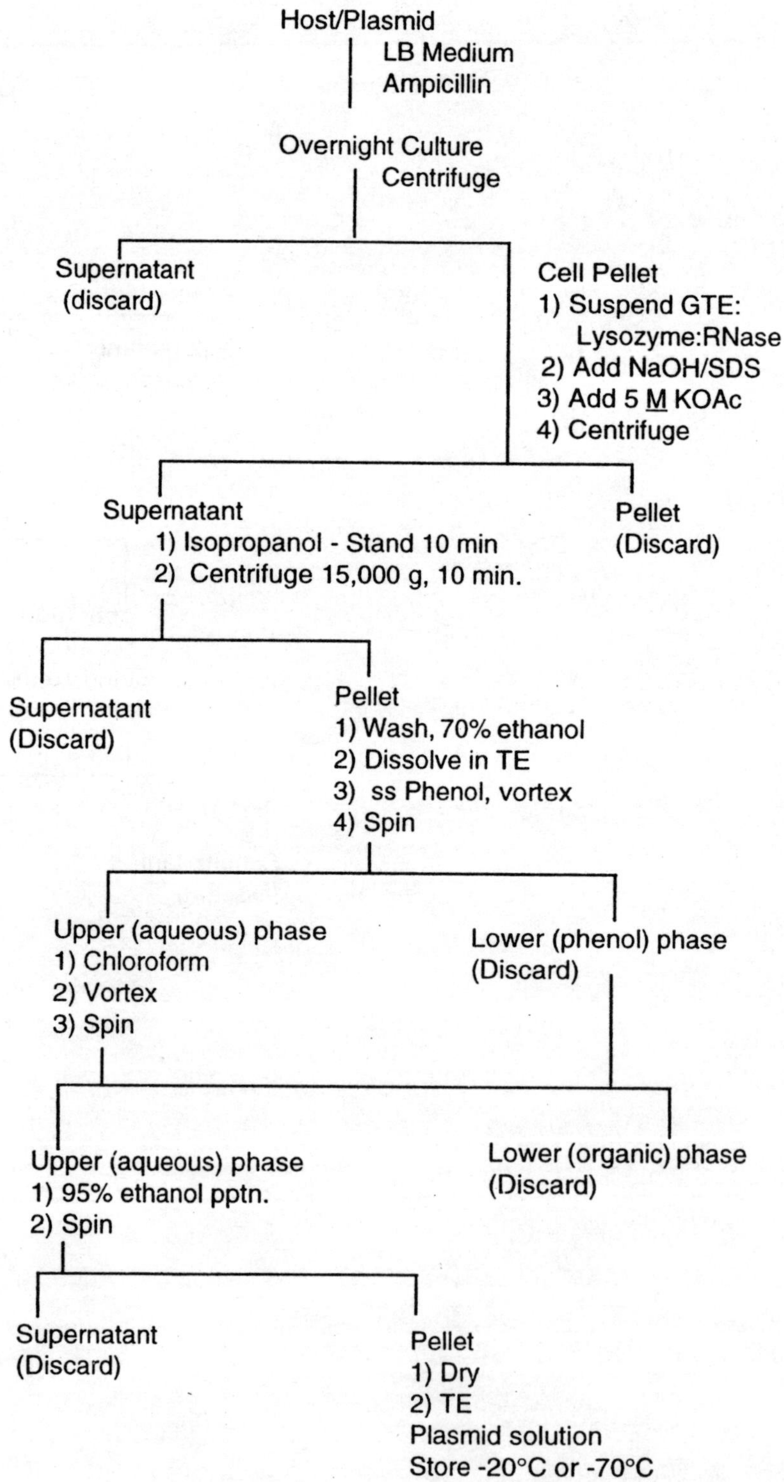

Materials

Plasmid Miniprep:

Host strain with plasmid
  (DH11S/pTZ18U)
LB Broth
Ampicillin (50 mg/ml) - Store solution at -20°C
Microfuge Tubes
Gloves
TELT Solution (100 ml)
  50 mM Tris-HCl, pH 8 (0.79 gm)
  62.5 mM Na_2EDTA (2.33 gm)
  2.5 M LiCl (10.6 gm)
  4 % (vol/vol) Triton X-100
1:1 (vol/vol) Phenol:Chloroform
RNase A
95% Ethanol
TE Buffer (100 ml)
  10 mM Tris (0.12 gm)
  1 mM EDTA, pH 8.0 (0.037 gm)

Goggles
Microfuge
Pipetors

Large Scale Prep:

Host/Plasmid (DH11S/pTZ18U)
LB Broth
Centrifuge Tubes (50 ml)
Microcentrifuge Tubes
Gloves (latex)
GTE Buffer (100 ml)
  50 mM Glucose (0.9 gm)
  25 mM Tris·Cl pH 8.0 (0.38 gm)
  10 mM EDTA (0.037 gm)
  Autoclave
Lysozyme
RNase A (10 mg/ml stock)
Lysis Solution
  .2 N NaOH
  1% (w/v) Sodium Dodecyl Sulfate (SDS)
5 M Potassium Acetate pH 4.8
  29.5 ml Glacial Acetic Acid
  Adjust pH to 4.8 with conc. KOH
  Adjust volume to 100 ml with ddH_20
Isopropanol
70% Ethanol

Materials, continued

Large Scale Prep, continued:

ss-Phenol
Chloroform
3 M Sodium Acetate
95% Ethanol
TE Buffer
Ampicillin (50 mg/ml)

Centrifuge (High Speed)
Microfuge
Pipetors

Gel Analysis of Plasmid:

Goggles
Gloves (latex)
Agarose
TBE Buffer pH 8.3 (10X)
  0.89 M Tris base (108 gm)
  0.89 M Boric Acid (55 gm)
  0.02 M $Na_2EDTA \cdot 2H_20$ (9.3 gm)
  ddH_20 to 1 liter
Tracking Dye/Loading Buffer
  40% (w/v) sucrose or glycerol
  0.25% bromophenol blue
  TBE (1x)
Methylene Blue Plus™
Ethidium Bromide (10 mg/ml)
Polaroid Film 667

Equipment:

UV white light
Transilluminator or hand held UV lamp (choice depends on stain)
Electrophoresis Cell
Power Supply
Microwave
Pipetors
Camera

Reagents for plasmid isolation are available, packaged as a kit, from EDVOTEK, Inc. Cat. #202: Mini-prep Isolation of Plasmid DNA. Telephone: 1-800-EDVOTEK; 1-301-251-5990

Method

Plasmid Miniprep:

1. Prepare a small (1-1.5 ml) overnight culture of Host/Plasmid such as *E. coli* strain DH11S/pTZ18U in LB medium. pTZ18U is a pUC derived plasmid which contains the gene for ampicillin resistance. When using systems containing plasmids of this type the growth medium is supplemented with 50 μg/ml ampicillin (use 1 μl of the stock solution above per ml of medium). Other plasmid-containing systems should be treated according to their specific selection characteristics.

2. Spin in the microfuge for 1 minute at high/maximum speed to pellet the cells.

3. Pour off the supernatant and resuspend the cells in 100 μl of TELT solution.

4. Add 100 μl of 1:1 phenol:chloroform. Shake or vortex vigorously.

5. Spin in the microfuge for 1 minute. Transfer the upper aqueous layer to a fresh 1.5 ml microfuge tube containing an equal volume of 95% ethanol.

6. Allow to stand for 10 minutes and spin in the microfuge at maximum speed for 10 minutes.

7. Pour off the supernatant, dry the pellet, and dissolve the pellet in 25 μl of TE buffer. The pellet may be dried by inverting over a paper towel and gently tapping, by directing a stream of air over the pellet, or by applying a vacuum to the tube. In the latter procedures, care must be taken not to lose the pellet.

8. If it is necessary to remove contaminating RNA, the sample can be digested with RNase A.

RNase A Digestion: Prepare a stock solution containing 10 mg/ml of RNase A. Heat in a boiling water bath for ten minutes to inactivate any DNase activity. Cool slowly. Store at -20°C. Add to a plasmid preparation to a concentration of 10-20 μg/ml and incubate for 30 minutes at 37°C. Remove protein by 1:1 phenol:chloroform extraction followed by ethanol precipitation, as done previously. Centrifuge and dissolve the pellet in TE buffer or sterile ddH_2O.

9. Plasmid DNA can be digested with restriction enzymes. Freeze sample until ready for digestion. RNA will not interfere with the digestion.

10. Alternatively prepare plasmid sample for electrophoresis. Add 5 μl of 10x gel load to a 50 μl plasmid sample and load the gel.

Method, continued

Large Scale Prep:

1. Prepare 25-40 ml of an overnight culture of Host/Plasmid (DH11S/pTZ18U) in LB medium.

2. Spin the suspension in a 50 ml tube at 8000 x g for 10 minutes at 4°C to pellet the cells.

3. Pour off the supernatant and resuspend the cells in 2 ml GTE buffer containing 20 mg lysozyme and 10 μg RNase A. Incubate at room temperature for 10 minutes. The lysozyme catalyzes the degradation of the bacterial cell wall; the RNAase is present to catalyze degradation of RNA; low temperature is used to minimize action of DNases during the extraction.

4. Add 4 ml of NaOH/SDS lysis solution and mix by gentle rocking. The mixture lyses the cells, the high pH denatures the double strand DNA (dsDNA) to single strand DNA (ssDNA), and the sodium dodecyl sulfate (SDS) complexes with proteins, denaturing them. Care must be taken to accomplish thorough mixing without undue agitation. The concern is that the large chromosomal DNA not be fragmented by shearing action caused by agitation. Were this to occur, contamination of the final product with sheared chromosomal DNA will result. One approach is to simply rock the tube gently seventy-five to eighty times. A marked viscosity change should be noted as the cells are lysed and the DNA released. Incubate in an ice bath for 10 minutes.

5. Add 4 ml of ice cold 5 M potassium acetate (pH 4.8), mix as before and incubate on ice for ten minutes. The potassium ions interact with the SDS protein/chromosomal complexes to form a precipitate, and the lower pH of this solution neutralizes the high pH caused by the lysis solution. This allows the plasmid to renature to dsDNA.

6. Centrifuge at 15,000 x g for 10 minutes. Carefully decant and save the supernatant to a clean 50 ml centrifuge tube.

7. Estimate the supernatant volume and add an equal volume of isopropanol. DNA is insoluble in this 50% (v/v) isopropanol solution but RNA retains a significant solubility. This aids in removing some RNA that may be contaminating the mixture. Instead of isopropanol, a double volume of 95% ethanol may be used. DNA and RNA are insoluble in this mixture.

8. Allow the mixture to stand at room temperature for 30 minutes to complete precipitation of DNA. Centrifuge at 15,000 x g for 10 minutes.

Method, continued

Large Scale Prep, continued:

9. Carefully discard the supernatant taking care not to lose any of the pellet. Suspend the pellet in 1 ml of 70% ethanol. This treatment serves to wash salts from the precipitate.

10. Transfer the suspension to a 1.5 ml microfuge tube and spin 3-5 minutes at maximum speed.

11. Pour off the supernatant and dissolve the pellet in 200 µl of TE buffer.

12. Add an equal volume of buffer-saturated-phenol (ss-phenol). CAUTION: Use gloves and goggles to safeguard against contact with the phenol which can damage tissue.

    ss-Phenol is prepared by mixing pure phenol with two or three volumes of TE buffer in a separatory funnel. The mixture is shaken vigorously and allowed to stand until the layers separate. The lower phenol layer is drained off. This process should be repeated several times with fresh buffer to completely neutralize the phenol (check pH of buffer with pH meter, not pH paper). The resulting ss-phenol is aliquoted and stored at -20°C.

    Saturating the phenol with buffer prevents "dehydration" of the aqueous phase by the phenol and loss of volume during the extraction. The phenol denatures contaminating proteins causing them to become insoluble. Upon centrifugation, they will accumulate at the interface of the organic (phenol) lower layer and the upper aqueous layer containing the DNA. Vigorously shake or vortex the tube to thoroughly mix the layers.

13. Centrifuge the tube for three minutes at high speed to effect separation of the layers.

14. Carefully, with a capillary-tipped pipet, avoiding any precipitate at the interface, remove the upper layer and transfer it to a clean microfuge tube. It is the usual practice not to attempt to save all the aqueous phase but to sacrifice some in order to avoid removing contaminating protein at the interface. If desired, the phenol extraction may be repeated, depending on the quality of the DNA desired.

15. To the aqueous phase containing the DNA, add an equal volume of chloroform, vortex or mix vigorously and spin in the microfuge at maximum speed to separate the layers. Remove and save the upper, aqueous layer as before.

Method, continued

Large Scale Prep, continued:

16. Add one tenth volume of 3 M sodium acetate and two volumes of 95% ethanol. Place in the freezer at -20°C overnight to precipitate the DNA.
17. Centrifuge at high speed for 3-5 minutes to pellet the DNA.
18. Pour off the supernatant and dry the pellet to remove traces of ethanol which inhibits enzymes used in later exercises. The pellet may be dried by inverting over a paper towel and gently tapping, by directing a stream of air over the pellet, or by applying a vacuum to the tube. In the latter procedures, care must be taken not to lose the pellet.
19. Dissolve the pellet in 50 μl TE buffer and store at -20°C.

Result

After extraction of DNA, the success of the exercise is assayed by performing agarose gel electrophoresis on a portion of the extract. Plasmids typically exist as several forms: CCC (covalently closed circular) which is unnicked, supercoiled, compact, and runs ahead of other forms; OC (open circular) which has at least one single strand nick so that the ds DNA is allowed to relax from the supercoiled form; linear, in which nicks occur in both strands in the same vicinity so that the circular form opens. The conformation of each of these forms varies so that their electrophoretic mobility differs leading to three bands in the gel upon detection. It is also possible that dimeric and higher multimeric forms may exist which further complicates the analysis. When the gel is run using TBE buffer, the mobility decreases in the order CCC, OC, linear. Submit samples to electrophoresis as described in Unit 1, Module 1. In this experiment, the amount of sample to load for electrohoresis is 10 microliters for staining with ethidium bromide.

UNIT 1: DNA ISOLATION

MODULE 3: PROKARYOTIC CHROMOSOMAL DNA

E. Robert Powell
Jack G. Chirikjian

- Introduction
- Safety Guidelines
- Experimental Outline
- Materials
- Pre-lab Preparation
- Method
- Results

Introduction

The isolation of high molecular weight chromosomal DNA is often the first step in molecular cloning. The chromosomal DNA of *E. coli* is a large circular molecule containing approximately 3,000,000 base pairs. The DNA is attached to the plasma membrane at several points. Large DNA is very sensitive to mechanical shear which causes random breaks in phosphate bonds of the molecule. However, if extraction is performed carefully, large fragments of chromosomal DNA can be obtained with an average fragment length of 100,000 to 200,000 base pairs. Since the average length of a gene is about 2,000 base pairs, there is a high probability that genes of interest will remain unbroken in one of the fragments of DNA. In subsequent steps, specific genes can be cloned.

The objective of this experiment is to isolate chromosomal DNA with a minimum of breaks. The resuspended cells are first mixed with ethylenediamine tetraacetic acid (EDTA). EDTA forms complexes (chelates) with several kinds of metal ions. Divalent metal cations, such as Mg^{+2} are required cofactors by the majority of DNases. The DNA being extracted is protected from DNase degradation since the complexed Mg^{+2} cannot be utilized by the enzyme. The addition of the ionic detergent sarkosyl dissolves the cell membrane and denatures many proteins. RNase is also present to degrade high molecular weight RNA. The proteolytic enzyme (protease) is added to the cell lysate in order to digest proteins that are free in solution or bound to the DNA. RNases and proteases are exceptionally stable enzymes. They remain active in the presence of denaturing detergents such as sarkosyl and at high temperatures. Eventually, the RNase will be degraded by the protease. However, it is in high enough concentration so that most of the RNA is degraded before significant proteolytic inactivation occurs. In the presence of salts, DNA and RNA precipitate from solutions containing high percentages of isopropanol or ethanol. Smaller molecules such as sugars and amino acids remain in solution.

In this experiment, the aqueous cell lysis solution is overlaid with cold isopropanol. High molecular weight DNA precipitates onto a glass rod when it is mixed at the interface of the two liquids. Due to its size and abundance, chromosomal DNA forms a viscous, clotted mass that is easily collected. This process is known as spooling. Since the RNA has been degraded by the RNase treatment, it is too short to cling to the rod. The spooled DNA is redissolved in buffer and analyzed by electrophoresis.

Safety Guidelines

This experiment is designed for staining of DNA with Methylene Blue Plus™ stain after electrophoresis. Methylene Blue Plus™ is formulated to provide optimal sensitivity for DNA visualization. Optimal visualization is obtained by using a Visible Light Gel Visualization system. As with any biological stain, care should be taken when handling solutions or gels containing methylene blue.

Gloves and goggles should be worn when handling methylene blue staining reagents, and worn routinely throughout the experiment as good laboratory practice.

Experimental Outline

Extract DNA from cells

↓

Spool DNA

↓

Redissolve DNA

↓

Prepare DNA for Electrophoresis

↓

Conduct Electrophoresis

Materials

This module utilizes EDVOTEK experiment #203, Isolation of *E. coli* Chromosomal DNA. The following components are included in the kit:

A. Buffer for cell resuspension
B. EDTA Buffer, 0.25 M, pH 8.0
C. RNase A (DNase-free)
D. Sarkosyl, 15%
E. Protease
F. NaCl solution, 5 M
G. Buffer concentrate for RNase, Protease and DNA resuspension (Tris-HCl, EDTA, pH 7.5)

1 Vial of DNA extraction LyphoCells™ (freeze-dried)
1 Tube 10x Gel Loading Solution
1 Tube Practice Gel Loading Solution
12 Tubes (large plastic)
10 Glass rods
1 Bottle of UltraSpec-Agarose™ powder (2.5 g)
1 Bottle of 50x concentrated electrophoresis buffer
1 Bottle of concentrated Methylene Blue Plus™ stain
1 1 ml pipet
1 100 ml plastic graduated cylinder
25 Microtipped transfer pipets

UltraSpec-Agarose™, LyphoCells™ and Methylene Blue Plus™ are trademarks of EDVOTEK, Inc.

Requirements

- Horizontal gel electrophoresis apparatus
- D.C. power supply
- Automatic micropipets with tips
- Water bath (45°C)
- Recommended: DNA visualization system
  (white light with Methylene Blue Plus™ staining)
- 50 ml beakers or flasks
- 250 ml beakers or flasks
- 5 and 10 ml pipets
- Pipet pumps
- Hot gloves
- Test tubes for DNA resuspension
- 91-100% isopropanol
  (70% isopropyl rubbing alcohol can be substituted - see Pre-Lab Instructions)

Pre-lab Preparation

Resuspension of DNA Extraction LyphoCells (A)

1. Add 8 ml of buffer for cell resuspension (A) to the vial of DNA Extraction LyphoCells.
2. Mix until all the material of the vial is resuspended. The suspension of cells will look turbid.
3. Transfer the entire contents of the vial to the rest of Buffer A (approximately 44 ml) in a beaker. Gently mix.
4. Aliquot 5 ml of resuspended cells to each of 10 large plastic tubes.
5. Cap the tubes and place them on ice or in the refrigerator until the laboratory activity.

Dilution of Buffer (G) Concentrate

Tris buffer (diluted buffer G) is used for preparation of RNase and protease solutions. The Tris buffer is also used by the students for DNA resuspension.

1. In a small beaker or flask, add all of the concentrated Tris buffer in bottle G to 36 ml of distilled water. Mix.
2. Store on ice or in the refrigerator.

Preparation of RNase (C) Solution
(RNase in Tris Buffer)

1. Add 6 ml of Tris buffer (diluted buffer G) to a large plastic tube.
2. With transfer pipet, add all concentrated RNase (C) to Tris buffer.
3. Rinse pipet in buffer to remove any residual RNase solution which might be left in the pipet.
4. Label the tube "RNase Solution" and store on ice or in the refrigerator.

Pre-lab Preparation, continued

Preparation of Protease (E) Solution
(Protease in Tris Buffer)

1. Add 12 ml of Tris buffer (diluted buffer G) to a large plastic tube.

2. Add all the powdered protease (E) to the Tris buffer. Cap and mix by inverting.

3. Label the tube "Protease Solution" and store on ice or in the refrigerator.

Preparation of Isopropanol:

1. Place the isopropanol in a flask in the freezer or on ice so it is thoroughly chilled before students spool the DNA.

   A. If you are using 91-100% isopropanol, place 100 ml on ice.

   B. If you are using 70% isopropyl rubbing alcohol, place 200 ml on ice.

Other Reagents:

The following kit components are ready to use as provided.

EDTA Buffer (B),
Sarkosyl Solution (D)
5 M NaCl Solution (F)

Method

Extracting DNA from Cells

1. Put your initials or group number on a tube containing 5 ml of resuspended cells.
2. Add 2 ml of EDTA Buffer (B) to the cell suspension.
3. Add 0.5 ml of RNase Solution. Tightly cap and mix by inverting the tube several times.
4. Let the cells incubate for 5 minutes at room temperature.
5. Add 0.5 ml of the Sarkosyl Solution (D).
6. Add 1 ml of Protease Solution. Tightly cap and gently mix by slowly inverting the tube 3 times.
7. Incubate the tube for 20 minutes in a 45°C water bath.
8. Add 0.6 ml of 5 M NaCl Solution (F). Mix by inverting.
9. Slowly pour the viscous DNA solution into a clean 50 ml beaker.

Spooling DNA

1 Carefully overlay the viscous DNA solution with the specified volume of ice cold isopropanol. Let the isopropanol slowly stream down the inside wall of the beaker or tube. Isopropanol is less dense than water so it will be the upper layer.

2. Submerge the end of a glass rod just below the interface of the isopropanol and the aqueous DNA solution.
3. Swirl rod several times in a circular motion to spool out DNA.
4. Remove rod to see if the precipitate is being collected. Precipitate will appear semi-transparent and gelatinous in texture.
5. Continue swirling rod to collect DNA from solution. Allow end of rod to occasionally touch the bottom of the beaker or tube.
6. Note the appearance of the spooled DNA. As the DNA adheres to the rod, its initial gelatinous texture (as it appeared in step 5) will become more compact and fibrous in appearance.

Method, continued

Re-dissolving Spooled DNA

1. Remove the rod from the beaker or tube, making sure to let the excess isopropanol drip off.

2. Add 2 ml of Tris Buffer (diluted buffer G) to a clean test tube.

3. Submerge the coated end of the rod into the buffer.

4. Twirl the rod several times to dislodge some of the DNA.

5. Cover the test tube, with the rod still inside, with plastic wrap, parafilm or foil to prevent evaporation.

6. Put your initials or group number on the tube.

7. Allow the DNA to rehydrate at room temperature. High molecular weight DNA can take several days to completely rehydrate and dissolve. Check the tube during your next lab period to see if some of the DNA precipitate has dissolved.

Preparing the DNA for Analysis by Electrophoresis

DNA can be analyzed by electrophoresis at a later time as determined by your instructor. To prepare the DNA for electrophoresis:

1. Transfer 0.3 ml of the dissolved DNA to a fresh test tube.

   Do not be concerned if the tube contains DNA that has remained undissolved.

2. Add one-tenth the volume (30 µl) of 10x Gel Loading Solution. (If you are using a transfer pipet, add 1-2 drops)

3. Mix by tapping.

4. The sample is ready for electrophoresis. Load 40 µl of the prepared sample on an agarose gel for electrophoresis.

Method, continued

Agarose Gel Electrophoresis

This experiment requires 0.8% agarose gel. Samples from various lab groups will be pooled for analysis on a shared classroom gel. Follow the instructions provided in Unit 1, Module 1 for preparing and running the agarose gel.

Loading DNA Samples

1. Each group should load 40 µl of the prepared DNA sample in a well of the agarose gel.
2. Remember to note the well in which each group loaded its sample.

Results

The figure to the left is an idealized schematic showing the relative mobility of the chromosomal DNA samples. The idealized schematic shows the relative positions of the bands, but are not depicted to scale.

Lane 1: Intact Chromosomal DNA

In this example, a band near the edge of the well may sometimes be observed. This extra band can be due to residual bound proteins that retard migration of the DNA into the gel.

Lane 3: Example of badly sheared chromosomal DNA

UNIT 1: DNA ISOLATION

MODULE 4:
EUKARYOTIC CHROMOSOMAL DNA: CALF THYMUS

John S. Boyle

* Introduction
* Safety Guidelines
* Experimental Outline
* Materials
* Pre-lab Preparation
* Method
* Results

Introduction

The objective of this module is to isolate chromosomal DNA from calf thymus cells by the process of ethanol precipitation. The initial step in the genetic manipulations involved in biotechnology is the isolation of DNA. In eukaryotic cells, chromosomal DNA is found within the nucleus. It does not exist as a free molecule in the nucleus, but rather as a complex association of DNA, RNA, and proteins. Negatively charged DNA is associated with numerous, positively charged, basic proteins. These include enzymes for replication, transcription, recombination, and repair as well as proteins for unwinding, supercoiling, and packaging. Protocols for DNA isolation require measures to remove the RNA and these proteins.

The following steps are employed to isolate DNA from eukaryotic cells:

1. Disruption of cell membranes by homogenization.

2. Isolation of the nuclei by centrifugation.

3. Disruption of nuclear membranes and denaturation of nucleoproteins by the action of detergents.

4. Precipitation and removal of the denatured proteins.

5. Precipitation and removal of DNA.

The reagents listed below are employed in this protocol to accomplish the five steps.

EDTA (ethylenediaminetetraacetic acid) is a chelating agent that removes divalent cations, such as Mg++ and Ca++. The removal of these ions weakens plasma and nuclear membranes, making them easier to lyse. The removal of these ions also inhibits DNA digestion by DNase enzymes released from homogenized lysosomes.

Sodium Dodecyl Sulfate (SDS) is a detergent which emulsifies the plasma and nuclear membranes promoting lysis. SDS also denatures protein.

Concentrated NaCl and Sodium Citrate solutions increase the ion concentration of the reaction mixture. These ions disrupt ionic bonds between DNA and protein. Concentrated NaCl also decreases the solubility of DNA in ethanol, allowing precipitation.

Buffered Sucrose solutions decrease the shearing of the long, fragile, DNA fibers.

Introduction, continued

95% Ethanol is used to precipitate DNA and remove it from solution in the aqueous phase. Salted DNA is insoluble in 95% ethanol and can be removed by spooling on a glass stirring rod. Increased purification may be obtained by using organic solvents, phenol and chloroform. These solvents further denature and dissolve protein, leaving DNA in the aqueous phase. However, they introduce increased safety risk to the laboratory and must be used with caution.

Care must be taken throughout all steps of DNA isolation to avoid contamination by DNases. These enzymes can be present on laboratory glassware. Latex gloves should be worn during the laboratory. All glassware should be cleaned.

Safety Guidelines

Safety glasses and gloves should be worn when performing this experimental module. Biological and other material should be disposed of in specially labelled containers and incinerated. No food or drink should be used in the laboratory.

Experimental Outline

2 ml well mixed suspended nuclei — 1 ml EDTA

Mix gently and let stand for 5 minutes

Add 0.3 ml SDS

Mix very gently, let stand 1 minute

Slowly add 0.3 ml 2M NaCl

Slowly add 2 ml 95% ethanol

Spool DNA with a glass stirring rod

Materials

Calf thymus - (results vary depending on freshness of thymus)
Knife for dicing thymus
Blender
Table top centrifuge
Centrifuge tubes- enough for one per student group
Cheesecloth
500 ml beaker

Pasteur pipets and bulbs or pumps
Small test tubes
Thin glass stirring rods
5 ml pipets
Pipetmen or bulbs for 5 ml pipets

Note: All glassware must be clean. Autoclaving or boiling is recommended.

Pre-lab Preparation

Allow 25 minutes to complete this preparation.

1. Dice 10 g of fresh calf thymus for a class of 30-50 students.
2. Immerse diced thymus in isotonic sucrose buffer in a blender and homogenize completely.
3. Strain homogenate through four layers of cheesecloth to remove clumps of gristle. Collect in a 500 ml beaker.
4. Pour homogenate equally into centrifuge tubes and spin for 10 min. at a medium speed (500 x g) in a table top centrifuge.
5. Pour off suspension and resuspend the pellet of nuclei in buffered sucrose.
6. Dispense tubes of suspended nuclei to student groups.

Preparation of Solutions

Sucrose Buffer

57 g sucrose
3.1 g $MgCl_2 \cdot 6H_2O$
0.6 g Tris-HCl
Distilled water to make 500 ml
Adjust pH to 7.5 with 0.1 N HCl
Store in refrigerator, lasts about one week

EDTA

0.72 g disodium EDTA$\cdot H_20$
250 ml distilled water
Adjust pH to 7.5 with 0.1N NaOH to dissolve EDTA
Keeps indefinitely at room temperature

SDS

25 g sodium dodecyl sulfate (sodium lauryl sulfate is the same)
250 ml distilled water
Keeps indefinitely at room temperature

2M NaCl

29.2 g sodium chloride
Distilled water to make 250 ml
Keeps indefinitely at room temperature

Method

1. Transfer 2 ml well mixed suspended nuclei to small test tube.

2. Add 1 ml of .28% EDTA, a chelating agent, to your tube and mix gently by holding the tube between the thumb and index finger of your left hand and tapping gently with your right index finger. Let stand for 5 minutes. EDTA removes divalent cations, Mg++, and Ca++, from solution and weakens the nuclear membrane. The removal of these ions also inhibits the activity of DNases, protecting the DNA from digestion.

3. Add 0.3 ml of SDS solution, mix very gently as above, and let stand for 1 minute. SDS (sodium dodecyl sulfate) is a detergent. It dissolves the nuclear membrane and denatures proteins associated with DNA, including DNases. The DNA is now fragile, so the following procedure should be done gently to avoid shearing the long strands.

4. Slowly add 0.3 ml of 2M NaCl to tube. Pause after each drop and mix gently. The 2M NaCl will interrupt the ionic bonding of protein to DNA. The addition of a strong NaCl solution will also render DNA insoluble in 95% ethanol in the next step.

5. Draw 2 ml of 95% ethanol into a 5 ml pipet. Slowly eject the ethanol down the side of your test tube, forming a layer of ethanol over the aqueous solution of DNA and protein.

6. Using a long, thin, sterile stirring rod, penetrate the boundary between the layers. Make slow, circular, stirring motions with the rod and spool the protein-free DNA by twirling the rod between two fingers. Continue swirling until you have a substantial clump of DNA.

7. If you wish to save the DNA, transfer the rod, or the tip of it, to a tube of ethanol and seal it.

8. If desired, the quantity of DNA isolated may be estimated by spectrophotometric analysis.

Results

Physical intactness of the extracted DNA can be assessed by analysis using 0.8% agarose gel electrophoresis as outlined in Unit 1, Module 1. Results obtained will be very similar to those shown in that module.

UNIT 2

Restriction-Modification Systems

UNIT 2: RESTRICTION-MODIFICATION SYSTEMS

MODULE 5: RESTRICTION ENZYME DIGESTION AND SIZE DETERMINATION OF DNA

Harley Mortensen

* Introduction
* Safety Guidelines
* Materials
* Method
* Results

Introduction

The objectives of this module are to demonstrate the activity of restriction endonucleases and to analyze the results of a restriction endonuclease digestion using agarose gel electrophoresis.

Three examples of Type II restriction endonucleases (abbreviated RE) will be studied. Type II RE recognize and "cut" (catalyze the hydrolysis of DNA) within palindromic sites (sequences which read the same as their inverse complement). An example is *Eco* RI which recognizes the sequence 5'-G^AATTC. Since the cut site is between the G and A bases, this will lead to fragments having overlapping ("sticky" or "cohesive") ends.

Before Digestion	After Digestion	
5'—GAATTC——3'	5'—G	AATTC——3'
3'—CTTAAG——5'	3'—CTTAA	G——5'

Some RE cut symmetrically (exactly in the center of the recognition site) and generate "blunt" ends. An estimate of the number of cut sites within a given piece of DNA is determined by $N/4^n$ where N is the number of base pairs in the DNA target piece and n is the number of bases in the recognition sequence. Since *Eco* RI recognizes a six-base sequence, 4^6=4,096. If the 4 bases of DNA are randomly distributed in the molecule, *Eco* RI will make one cut for every 4096 base pairs. Therefore, DNA such as Lambda, which has approximately N = 48,500 base pairs, would theoretically be cut into 11 or 12 fragments by *Eco* RI.

In this exercise Lambda DNA will be incubated with the restriction enzymes *Hind* III, *Eco* RI, and *Bam* HI. The recognition sequences for *Bam* HI and for *Hind* III are G^GATTC and A^AGCTT, respectively. Subsequent gel electrophoresis of the individual reaction mixtures will illustrate the different fragmentation patterns produced by the enzymes. The linear map of Lambda (Figure 1) shows the location of sites of action by the three restriction endonucleases.

For detection by ethidium bromide, it is desirable to have approximately 100-200 ng of DNA per gel band. As an example, since *Hind* III generates six or seven bands on a typical gel, a total of approximately 1.5 µg (1500 ng) of DNA should be used. Lambda DNA can be purchased in a concentration of about 0.5 µg/µl.

Size Determination of DNA

It is frequently necessary to determine the size of a particular piece of DNA. One convenient way to accomplish this is through the gel procedures that are commonly applied to proteins. In these instances DNA, whose size is to be determined, is run out on a gel. A molecular weight standard is run on a separate lane of the same

Introduction, continued

gel. The standard's fragments are all of known size and can be used to construct a standard curve relating some function of size (log M.W., for example) versus migration distance in the gel. By measuring the migration distance of the unknown material, the size can be taken from the standard curve.

In protein work the gel used is polyacrylamide. However, the typical DNA fragment is much too large to be effectively separated using polyacrylamide. Agarose gels are used instead with adjustments in agarose percentage made in order to accommodate the possible range in size of DNA pieces that might be encountered. The table below shows some typical agarose gel percentages appropriate for various DNA sizes.

AGAROSE (%)	Effective Range of Resolution of Linear DNA Fragments (kb)
.5	30-1
.7	12-0.8
1.0	10-0.5
1.2	7-0.4
1.5	3-0.2

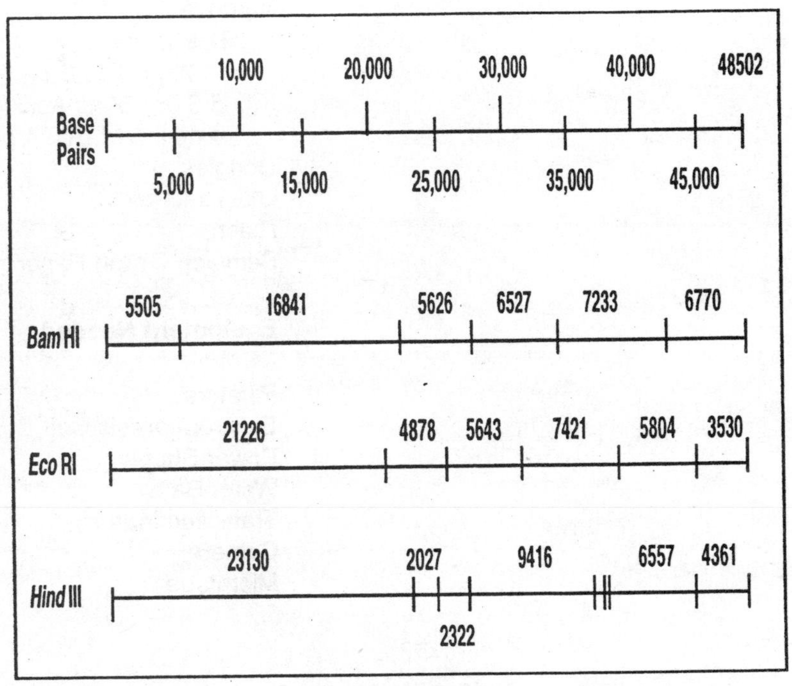

Fig. 1: Linear Map of Lambda DNA

Safety Guidelines

Use gloves when handling ethidium bromide. Ethidium bromide is a powerful mutagen. Use goggles or other proper eye protection when viewing gels through a transilluminator. Ultraviolet radiation from the transilluminator can cause tissue damage. For introductory classes, Methylene blue staining of DNA is recommended to minimize hazards associated with ethidium bromide and its disposal.

Materials

Lambda DNA (0.5 mg/ml in TE buffer)
Eco RI
Hind III
Bam HI
Core Buffers (obtained from enzyme supplier)
Microfuge Tubes
Pipet Tips
Tracking Dye
Ethidium Bromide stock (10 mg/ml) or
  Methylene Blue
Polaroid Film 667
Agarose
1 X TBE Buffer
    10.7 gm Tris Base (0.089 M)
    5.5 gm Boric Acid (0.089 M)
    4 ml 0.5 M EDTA pH 8.0 (0.008M)
Goggles
Gloves (latex)
Ruler
Semi-log Graph Paper

Equipment Needed

Pipetors
Electrophoresis Cell
Power Supply
Water Bath
Transilluminator
Camera
Microfuge

Method

Restriction Enzyme Digestion

1. Set up a restriction digestion as follows using three 1.5 ml microfuge tubes

	TUBE 1	TUBE 2	TUBE 3
Lambda DNA	3 µl	3 µl	3 µl
10X core buffer	2.5 µl	2.5 µl	2.5 µl
dd water	17.5 µl	17.5 µl	17.5 µl
Hind III	2 µl	0	0
Eco RI	0	2 µl	0
Bam HI	0	0	2 µl

NOTE: Core buffer usually supplied with purchases of enzymes.

Centrifuge briefly in a microfuge to mix.

2. Incubate the tubes for 45-60 minutes at 37°C.

3. Denature proteins by heating in a 75°C water bath for 5 minutes.

4. Prepare a 0.75% agarose gel in TBE (1x). Heat in a microwave until boiling begins. Stir and heat just to boiling. Cool to 50°C. Place the well comb in position and pour enough of the warm agarose to form a gel of the desired thickness. Allow to cool for 30-60 minutes.

5. Add 2 µl of tracking dye to each digest and load one gel lane with 10 µl and another with 15 µl of the *Hind* III digest. Do the same for the *Eco* RI and *Bam* HI digests.

6. Proceed with agarose gel electrophoresis as described in Unit 1, Module 1.

Results

Size Determination of DNA

The sizes, in base pairs (bp) of the *Hind* III fragments from digestion of lambda DNA are: 23,130; 9,416; 4,361; 2,322; 2,027; 564; 125. The 125 bp fragment runs off the gel and is not detected. Depending on the time of the run, the 564 bp fragment may also run off the gel.

Results, continued

1. Determine the migration distance in millimeters for each *Hind* III fragment in the most readable lane by measuring the distance from the bottom of the well to the middle of the fragment.

2. Using semi-log graph paper, construct a standard curve by plotting size (in base pairs) on the log axis vs migration distance in mm) on the linear axis. Connect the points by a smooth curve.

3. Choose one or more appropriate fragments from the *Eco* RI and *Bam* HI lanes and measure their migration distances.

4. Locate that distance on the linear axis of the standard curve and find the point where that value intersects the curve. The bp value which determines the other coordinate of the point is read off the log axis to obtain the base pair size of the fragment.

5. Compare the experimental value with the accepted value.

6. Calculate the percent error in the determination.

   Eco RI: 21,226; 7,421; 5,804; 5,643; 4,878; 3,530
   Bam HI: 16,841; 7,233; 6,770; 6,527; 5,626; 5,505

UNIT 2: RESTRICTION-MODIFICATION SYSTEMS

MODULE 6: ISOLATION OF RESTRICTION ENZYMES

Jack G. Chirikjian

* Introduction
* Safety Guidelines
* Materials
* Pre-lab Preparation
* Method
* Results

Introduction

Sequence-specific, or Type II, endonucleases are commonly described as restriction enzymes. In contrast with nonspecific or Type I enzymes, Type II endonucleases generate reproducible nucleotide fragments from specific DNAs. They cleave double-stranded DNA by hydrolyzing two phosphodiester bonds (one per strand) within defined nucleotide sequences. Over 1500 enzymes have been discovered since the first report by H. O. Smith and collaborators, with a variety of bacterial cells as their source.

The name of a restriction enzyme is derived from the genus and species of bacterium from which it is isolated. The first letter of the genus name and first two letters of the species are combined to form the enzyme name. This is followed by a strain designation if needed. In many instances, a bacterial strain contains more than one restriction endonuclease. When this occurs, each enzyme is assigned a Roman numeral. Thus, *Bam* HI was the first enzyme activity reported from *Bacillus amyloliquefaciens* strain H.

Structurally, most restriction enzymes reported in purified form are composed of two equal subunits with molecular weights of 20,000-25,000 or single polypeptides with molecular weights of 30,000-35,000. Enzyme activities can be differentiated from each other by their characteristic patterns of digestion of small viral DNAs. The DNA from bacteriophage Lambda is the most widely used substrate for screening restriction enzymes. Because it is often difficult to determine a characteristic pattern from a Lambda digest, smaller DNAs, such as the replicative form of bacteriophage ØX174 and of SV40 are also used as substrates. The resulting DNA digest is displayed on agarose or polyacrylamide gels and visualized by staining with Methylene Blue Plus™ or ethidium bromide.

A given sequence in DNA is recognized and can be cleaved by more than one restriction enzyme. The term "isoschizomers" describes a group of enzymes that recognize the same sequence in DNA. In such cases, the representative enzyme that is used most commonly is selected on the basis of ease of purification and stability in purified form. The sequences recognized by these enzymes are for the most part centrosymmetric "palindromic" sequences that are usually hexamers, pentamers, or tetramers. Several Type II enzymes recognize DNA at a specific site and hydrolyze phosphodiester bonds at a defined distance from that site. An example of this group of enzymes is *Bgl* I, which recognizes a sequence containing two groups of specified residues separated by completely unspecified residues - GCCNNNNNGGC; it therefore generates DNA fragments with variable end groups.

There is considerable diversity in the fragment termini produced in cleavage by Type II endonucleases that recognize and cleave within the same sequence. In some cases, the 5' extension may be

Introduction, continued

as short as two nucleotides or as long as five. Points of cleavage on each strand may be opposite each other; this results in blunt (square) ends. Several restriction endonucleases produce 3' extensions of two to four nucleotides. However, all Type II endonucleases produce fragments with a 5'-terminal phosphate and a 3'-terminal hydroxyl residue.

Enzymes in this family are amenable to purification by conventional isolation procedures. Phosphocellulose at nearly neutral pH is a widely used separation matrix for extracts that have been freed of cellular nucleic acids. At this stage of purification, short-term assays often make it possible to visualize fractions that contain enzymes in the presence of nonspecific exonucleases.

A variety of enzymes have been fractionated with affinity chromatography. This method takes advantage of biospecific interactions not offered by conventional fractionation methods. Enzymes fractionated by such procedures can be concentrated selectively and eluted by gradients containing substrates or salt. The advantages of affinity chromatography are speed of purification and often protection against denaturation during fractionation. A general procedure to purify restriction enzymes utilizes Cibacron blue. Cibacron blue F3GA is a blue dye that appears to have biospecific affinity for nucleotide-requiring enzymes. The matrix is commercially available from a variety of sources or can be prepared as described below. Before fractionation of enzymes on blue-CNBr-sepharose, any unbound ligand should be removed by salt washes and extraction with a hydrophobic solvent, such as dioxane.

Effect of Hydrophobic Reagents on Catalysis by Restriction Enzyme

Several reports have described apparent changes in specificity of restriction endonucleases in association with an altered reaction environment. Conditions that alter specificity have included changes in ionic concentration, in pH of the reaction buffer, and the introduction of glycerol into the reaction mixture.

Introduction, continued

Table 1: Partial list of restriction enzymes whose sequence recognition is altered by glycerol.

Enzymes that Display Altered Specificity†	
Ava I	*Hpa* I
Bam HI	*Pst* I
Bst I	*Sal* I
Eco RI	*Sst* I
Hae III	*Sst* II
Hha I	*Xba* I

†Reaction mixtures contained 10mM Tris (pH 8.5), 10 mM $MgCl_2$, 10mM-mercaptoethanol. To determine the effect of glycerol the same reaction mixtures were used with various concentration (11 to 61%) glycerol and DMSO with concentration of up to 17% and at least 25-fold enzyme excess. Incubations were carried out 12 hours. Three different DNA substrates (Lambda, SV40, ØX174) were tested. Change of DNA digestion patterns was noted with at least one DNA.

A change in *Bam* HI activity is detected during the purification procedure. The second activity is designated at *Bam* HI.1. A similar activity is displayed by *Eco* RI. Increasing the pH of the reaction from 7.0 to 9.0 in the absence of monovalent cations stimulates *Bam* HI.1. activity. Decreasing the ionic strength had a similar effect. Varying the $MgCl_2$ concentration from 5 to 20 mM had no effect on either *Bam* HI or *Bam* HI.1 activities. Neither activity was obtained when $MgCl_2$ was replaced by Zn^{2+}, Co^{2+} or Cu^{2+}.

Introduction of glycerol into the reaction mixture caused the greatest alteration of enzyme activity. Lambda DNA was incubated with *Bam* HI in the presence of glycerol at various concentrations (Fig 1, lanes 1-8). In the absence of glycerol, the usual *Bam* HI-DNA pattern was obtained. At 1% glycerol (lane 2), the *Bam* HI pattern of Lambda DNA was distinctly present, with several low-molecular weight DNA fragments. A progressive change in the DNA digestion pattern was observed when glycerol concentration was increased from 11 to 61% in the reaction mixtures (lanes 3-8). At high concentrations, a different DNA banding pattern was observed. Glycerol samples from various manufacturers led to the identical results. DMSO (lanes 14-18) could substitute for glycerol (lanes 6-8) to generate identical terminal-digest patterns.

As an additional exercise, the class can also investigate the effect of duration of incubation on the appearance of *Bam* HI.1 activity. Constant amounts of enzyme and DNA can be incubated with 17% DMSO (dimethylsulfoxide) for various periods (lanes 10-18). *Bam* HI.1 activity will be observed in as little as 20 minutes (lane 10), and

Introduction, continued

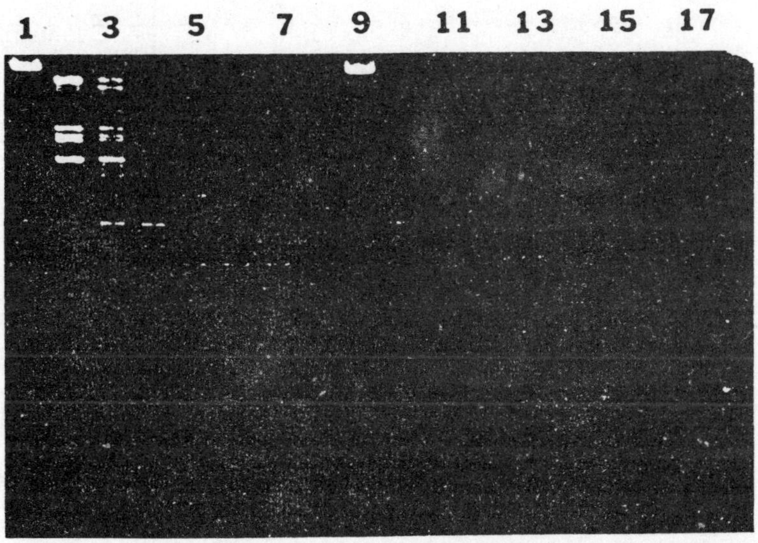

Figure 1: Effect of glycerol and M_2SO on *Bam* HI. Glycerol (lanes 1-8) and M_2SO were introduced into *Bam* HI standard reaction mixture containing glycerol at final concentration of 1 to 61% (v/v) in 10% increments incubated at 37°C for 1 hour. Lanes 10-18, reaction mixtures containing 17% (v/v) M_2SO with incubation time of 20-180 minutes in increments of 20 minutes.

the *Bam* HI pattern will be absent beyond 40 minutes (lane 11). Further incubation, to at least 180 minutes (lane 18), will not change the observed pattern of DNA fragments. The final DNA fragmentation patterns obtained with incubation with glycerol (lane 8) or DMSO (lane 18) will be essentially identical. The effects of ethylene glycol, ethanol, and dioxane on *Bam* HI activity is the same as glycerol and DMSO.

Safety Guidelines

Use gloves when handling ethidium bromide. Ethidium bromide is a powerful mutagen. Use goggles or other proper eye protection when viewing gels through a transilluminator. Ultraviolet radiation from a transilluminator can cause tissue damage. Methylene blue staining of DNA is recommended for introductory classes to minimize hazards associated with ethidium bromide and its disposal.

Experimental Outline

Cell Growth

- Overnight at 32-37°C
- Harvest cells and Freeze

Preparation of Bacterial Crude Extract

- Removal of Nucleic Acids
- Centrifugation at 10,000xg for 15 minutes

Purification of Restriction Enzymes

- Assay Fractions Using Lambda

Pooled partially Purified Enzyme

Determine Total Units

- Effect of pH
- Effect of Salt
- Effect of Glycerol

Materials

Cell-seed cultures for one or more (*Bam* HI, *Pst* I and *Xba* I)
DEAE-Cellulose
Media-Tryptone, yeast extract
Cybacron blue matrix
Glycerol and/or DMSO
Lambda DNA
Electrophoresis buffer
Agarose

Reagents for the experiment are available in kit form from EDVOTEK, Inc. Telephone 1-800-EDVOTEK.

Pre-lab Preparation

Cibacron blue sepharose can be prepared by using CNBr-activated sepharose or by activating sepharose with cyanogen bromide. CNBr-activated Sepharose can also be purchased commercially from Pharmacia, Inc.

In order to activate the sepharose 4B (Pharmacia), 100 g is suspended in 100 ml of distilled water and 100 mg of finely divided solid cyanogen bromide (CNBr) is added over 5 min. The pH is adjusted and maintained at 11 by addition of 8 N NaOH with stirring. After 10 min. at 20°C, 200 ml of cold distilled water is added. The suspension is washed and the activated sepharose is used. This has to be done in a well ventilated hood and should not be attempted by students.

CNBr-activated sepharose, either purchased or prepared, is washed with excess 1mM HCl, and Cibacron blue F3GA is coupled at pH 8.3 in 0.1M $NaHCO_3$ and 0.5 M NaCl at 4°C for 16 hours. If necessary, pH is maintained at 8.3 with 0.1N NaOH. The ratio of starting blue dye to swelled activated sepharose is 1 g to 10 ml. To block the remaining active groups, 1 M Tris·HCl pH 9.0 is added and incubated for 1 hour at 22°C. The matrix can be stored at 4°C in the presence of high NaCl and a bacteriocide and will remain stable for several months.

Method

Cell Growth

Bacterial cultures are available from ATCC (Rockville, MD).

1. Cells can be grown by inoculating the specific medium with a 2% overnight culture.

2. After growing for 16 hours, cells are harvested by centrifugation at 8000 x g, washed with 0.9% saline and frozen at -90°C.

3. Growth media and temperature conditions for the different strains are as follows: *Bacillus amyloliquefaciens. Providencia stuartii* 164 is grown on tryptone (11 g/l), yeast extract (22 g/l), glycerol (4 ml), K_2HPO_4 (8.9 g) or KH_2PO_4 (2.14 g) per liter and NaCl (5 μg/l) at 37°C; *Xanthomonas badrii* is grown on nutrient broth (8 g/l) at 32°C.

Method, continued

Analysis of DNA Endonucleases Activities by Gel Electrophoresis

Assays are performed in 50 µl volumes in either optimal buffers (details for each specific enzyme are indicated in the Table 1 legend).

1. Selection of the lambda DNA is based on obtaining a characteristic and recognizable fragmentation pattern.
2. Incubations (10 µl) of column fractions and assays to locate activity at 37°C for 1 hour or overnight.
3. Reactions are stopped by the addition of 5 µl of reaction stop solution (for routine assays, gels at a concentration of 0.8% agarose are prepared and electrophoresed in Tris (48.4 g/l), sodium acetate (16.4 g/l) and EDTA (7.4 g/l); adjusted to pH 7.8 with acetic acid and run for 3 hours at 6 volts/cm).
4. Gels are stained with Methylene Blue Plus™ or ethidium bromide, and DNA is visualized with short wavelength UV light source.
5. Photographs are obtained by the use of a red filter (Kodak 23A) and Polaroid type 57 film.

Unit Determination

Unit determination is performed with serial dilution of enzyme using a fixed amount of substrate. A unit of enzyme is defined as the minimum amount of enzyme which will totally digest 1 µg of Lambda DNA in 60 min at 37°C.

Preparation of Bacterial Crude Extracts

A typical preparation utilizes 15 g of pre-washed bacterial cells.

1. Cells are suspended in 35 ml of Buffer A (20 mM Tris-HCl pH 7.2, 1 mM Na_2EDTA, 7 mM β-mercaptoethanol and 10% glycerol).
2. Cells are at least 80% lysed by sonication using 30 second pulses for 8-10 minutes with intermittent on-off cycles.
3. The temperature of the slurry is maintained at 10°C through the lysing, and all procedures to follow are carried out at 4°C unless otherwise specified.

Method, continued

4. The cell extract is then separated by two sequential centrifugation steps: at 10,000 x g for 1 hour each. Only supernatant (approximately 40 ml), referred to as crude extract, is recovered and pellets discarded.

5. Nucleic acids are then removed by one of several procedures, e.g., batch elution from DEAE-cellulose under conditions where the enzyme is recovered, or by precipitation with streptomycin sulfate and PEG-6000. Since nucleic acids will bind to the column reducing the capacity of the matrix, acid was found to be a prerequisite.

6. The supernatant after the removal of nucleic acid is submitted to fractionation by Cibacron blue sepharose as described in the legends.

Results

Purification of Restriction Endonucleases

Cibacron blue sepharose is effective in the separation of restriction endonucleases from nucleic acid free bacterial extracts. For this experiment, we chose three enzymes: *Bam* HI, *Pst* I and *Xba* I, obtained from three different genera. For the initial experiments, a 1.5 ml column of blue-CNBr-sepharose is equilibrated with Buffer A. Columns were developed with a 20 ml salt gradient 0-0.8 M in Buffer A. Salt concentration at the peak of enzyme activity for these and other enzymes tested is summarized in Table 1. In most preparations, yields between 2,000 and 10,000 units are obtained.

Table 2: Salt Concentration at Peak of Endonuclease Activity

Organism	Enzyme	Salt Concentration (Molar) (Approximated)
Bacillus amyloliquefaciens	*Bam* HI	0.45 M
Providencia stuartii 164	*Pst* I	0.30 M
Xanthomanus badrii	*Xba* I	0.25 M

UNIT 2: RESTRICTION-MODIFICATION SYSTEMS

MODULE 7: RESTRICTION-MODIFICATION OF DNA

Peter B. Woodruff

- * Introduction
- * Safety Guidelines
- * Materials
- * Pre-lab Preparation
- * Method
- * Results

Introduction

The unprecedented advances made in molecular genetics over the past two decades owe a lot to the identification, isolation and use of restriction endonucleases. Discovery of these enzymes brought precision to the "cutting" of DNA sequences, almost overnight simplifying the analysis and manipulation of nucleic acids. Today they are such an integral part of DNA science that it is hard to imagine a molecular genetics lab without them.

Restriction

Although restriction enzymes are of bacterial origin, the adjective restriction refers to the observed host strain restriction exhibited towards many bacteriophages.

Recall that lytic phage release hundreds of progeny in the lysis or destruction of their host cell, which go on and repeat the process in surrounding cells. If dilute suspensions of phage are spread on a lawn of bacteria growing on agar in a Petri plate, their reproductive success may be monitored by observing the relative number of clear areas of plaques which result from these repeated rounds of cell lysis. This is recorded as the "efficiency of plating" or "eop", which is around 1 under nonrestricting conditions.

In the early 1950s, a future Nobel laureate quietly reported a "...non-hereditary host induced variation of bacterial viruses". Little did he realize that nearly twenty years later refinements on his observations would begin the biotechnology revolution.

Restriction Modification

In the classic example of restriction, studies by Arber and collaborators showed that phage which consistently grew well on lawns of *E. coli* strain K produced few plaques on *E. coli* strain B (ie: an eop of 10^{-4}). The phage were said to be restricted to strain K. What made this phenomenon all the more intriguing was the observation that phage isolated from those few stain B plaques which escaped restriction grew normally through repeated platings on strain B, yet formed few plaques on strain K.

The explanation for how genetically identical phage could exhibit strain specificity lies in the unique restriction-modification (R-M) enzyme systems of their hosts. By enzymatically flagging (modifying) their own DNA and degrading unflagged DNA on the spot, many prokaryotes avoid infection by unmodified phage DNA. On lawns of the restricting host, the occasional viral plaque (typically with an eop of between 10^{-3} and 10^{-5}) identifies a case where the invading phage DNA has become protectively modified in the host pattern before the host cell has a chance to degrade it. Within the host cell, protected phage produce progeny containing modified

Introduction, continued

DNA. Once released, these phage are free to infect and lyse surrounding cells of the same strain, producing the plaque.

Thus phage are restricted more or less to bacterial host strains which share the modification pattern of their DNA. In the rare instance where phage take on the modification pattern of a new host strain, they lose their ability to reproduce in the earlier host strain.

This type of phage containment system appears to confer considerable evolutionary advantage: bacteria from all taxonomic and ecological backgrounds possess R-M systems which apparently arose repeatedly and independently throughout the moneral kingdom. Of over 10,000 bacterial strains tested to date, more than one-quarter show evidence for R-M systems, and this is probably an underestimate. A majority of bacterial strains testing positive for R-M systems possess one or more; some even have six or seven. Though quite divergent in detail, they share many striking features.

Central to each of these systems are two antagonistic enzyme activities which share a common 4-8 nucleotide (nt) recognition site. One, a DNA methyltransferase (MTase), methylates one adenosine or cytosine of each strand of the recognition sequence, thereby conferring protection from cleavage by the other, a restriction endonuclease that causes double-strand breaks in unmethylated DNA.

The methylation reaction always uses S-Adenosyl-L-methionine (SAM) as the methyl donor. Each methylase produces its own characteristic pattern of methylation:

M·*Eco* RI	M·*Bam* HI
CH_3	CH_3
5'...GAATTC...3'	5'...GGATCC...3'
3'...CTTAAG...5'	3'...CCTAGG...5'
CH_3	CH_3

Only N6-methyladenine, N4-methylcytosine and 5-methylcytosine have been identified as methylase products. Their methyl groups project into the major groove of DNA without interfering with base pairing and apparently prevent cleavage by the "cognate" (same system) restriction endonuclease through steric interference.

Most Type II restriction endonucleases cleave unmodified DNA within or close to the recognition sequence, although a few cut at variable distances considerably removed from the recognition site. All require divalent cations, MG^{++} in particular, and Type I RE's show a preference for ATP or SAM.

Introduction, continued

Newly synthesized DNA is *hemimethylated*, with the old strand retaining its methyl group and the new one temporarily without. This hemimethylated DNA is the usual substrate for the DNA-methyltransferase in the cell, rendering it fully methylated again. The cell's restriction endonuclease system only cleaves DNA which is unmethylated at its recognition sites.

Restriction-modification systems have been grouped into four families (I, II, IIS, & III) based on their protein structure, recognition and cleavage sites, and cofactor requirements. In practice, only Type II enzymes are used in genetic engineering.

Restriction Endonucleases: The Cutting Edges of Biotechnology Research

Endonucleases are enzymes that hydrolyze phosphodiester bonds at internal locations within a polynucleotide chain. Exonucleases, by comparison, break phosphodiester bonds sequentially, starting at one free end of a polynucleotide and working towards the other. Restriction endonucleases cleave duplex DNA only at specific sites, producing double strand breaks. They generally cut to the 5' side of the phosphate, resulting in breaks which are amenable to the DNA ligase repair process, suggesting that such accidental nicking and repair may happen in nature.

For Type II enzymes, the cleavage site is generally within the recognition site. The recognition and cleavage sites are somewhat displaced in Type IIS and III systems, while Type I enzymes often cut over a thousand base pairs from their recognition site.

Nomenclature: R-M enzyme acronyms include the type of enzyme. R for restriction endonuclease, M for modification methylase, (omitted if the context is unambiguous), followed by an indication of its biological source and, if more than one R-M system are present in one strain or different substrains of the same strain, a roman numeral assigned to the particular system. Applying this to the endonuclease *Eco* RI we have:

enzyme designation if required	1st letter of Genus name	1st two letters of species name	strain, if important	number of enzyme
R	*Escherischia*	*coli*	R	I

The protein designation for an R-M enzyme often carries over to the short form for its gene, with modified order and case: for the *Bam* HI R-M system, the genes are *Bam* HIR *and Bam* HIM).

Introduction, continued

Properties of Type II Restriction Endonucleases

For molecular biology research purposes, Type II restriction endonucleases are the tools of choice.

The endonuclease *Eco* RI is fairly typical of Type II. It is derived from the *E. coli* RY 13 strain[1] and recognizes the unmethylated sequence:

5'...GAATTC...3' and cleaves it to form: 5'...G AATTC...3'
3'...CTTAAG...5' 3'...CTAA G...5'

This illustrates several points common to many Type II endonucleases. The symmetry seen in the recognition sequence is a form of palindrome, in that both strands read the same (GAATTC) when read in the 5' to 3' direction. DNA cut at both of the staggered cleavage sites possesses single stranded sticky ends. As their name suggests, under appropriate conditions these single stranded regions may reassociate, forming a DNA duplex but retaining the covalent breaks between G and A. DNA ligase will mend the breaks.

These apparently simple properties of restriction endonucleases and DNA ligase are central to one of the most widely applied and crucial approaches available to molecular biologists; the formation of chimeric DNA formed by ligating DNA from disparate sources.

One can make further generalizations about Type II restriction endonucleases from consideration of just a few (see Figure 1). Most restriction enzymes recognize palindromic sequences of 4-8 pairs, with Type II rarely going beyond 6bp. Many enzymes yield single-stranded sticky ends on cleavage; others cut completely across the DNA duplex to form blunt ends. Some sequences are degenerate. Others show only partial symmetry, hyphenation or no symmetry at all.

Cleavage by particular restriction endonucleases may result in any one of the following patterns:

1. Staggered, protruding 5' phosphates, as with *Eco* RI
2. Staggered, recessed 5' phosphates, as with 3'-OH protruding ends, as for *Bgl* I
3. Blunt ends, 5' phosphate, 3'-OH, as for *Hae* III.

---

[1] Unlike most nucleases, its code is on the strain's R plasmid.

Introduction, continued

```
            ↓                      ↓        ↓
5'...G AATT C...GG CC...GCCNNNN NGGC...A AGCT T...3'
     x          x          x            x
3'...C TTAA G...CC GG...CGGN NNNNCCG...T TGCA A...5'
           ↑      ↑        ↑                ↑
```

R·*Eco* RI	R·*Hae* III	R·*Bgl* I	R·*Hind* III
Escherischia coli RY13	*Haemophilus aegyptius*	*Bacillus globigii*	*Haemophilus influenza*

Figure 1 - Specificities of some restriction endonucleases: the recognition sequences are shown, while the arrows locate the cleavage sites. "N" refers to any nucleotide.

Enzymes from different systems which recognize the same sequence are grouped as isoschizomers. Of the well over 1500 restriction endonucleases that have been identified to date, over two hundred different recognition sequences have been identified.

In a complete digest, a restriction endonuclease will cut a particular DNA whenever its unmethylated recognition site occurs. As a rule of thumb the cleavage frequency = 4^n, where n is the number of base pairs in the recognition site. This assumes random distribution of equal numbers of A, T, G and C, Hence, if n=4 (eg, R·*Hae* III), the site will occur about every 256 bases, if n=6 (eg, R·*Eco* RI), statistically it will cleave once every 4096 nucleotides.

In general, the longer the molecule, the greater the probability that a given recognition site will occur.

Many more fragments result from a restriction endonuclease digest of chromosomal DNA than from comparable treatment of plasmid or viral DNA containing only a few thousand base pairs. The fragments produced after restriction digestion of bacterial or eukaryotic chromosomal material range so widely in size and are so numerous that they usually appear as a smear on subsequent agarose gel electrophoresis. In contrast, cutting plasmid or viral DNA with selected restriction endonucleases yields a limited number of bands each usually representing a discrete DNA fragment.

Closed circular DNA with a single restriction site will open up to form a single linear molecule after cleavage. A linear DNA molecule with a single site cleaves to form two fragments. The relative fragment size depends on the position of the restriction site. For instance, more or less equal sized fragments about half of the size of the original result if the site is near the center of the molecule.

Introduction, continued

For some purposes, reaction conditions such as reaction time can be controlled to produce partial cleavage, leaving some DNA molecules cut less completely than others.

The affinity of some DNases for their recognition sequences is significantly affected by the flanking sequences around their recognition site. The DNase probably responds to conformational differences in the DNA due to the differences in flanking sequence. If the effect is pronounced, fewer sites than expected will be cleaved and partials will result.

Sequence specificity may be perpetuated if the reaction conditions stray from optimal. Conditions which encourage perturbation and lead to so-called star activity, include high pH (8.5), low ionic strength, presence of organic solvents (eg: glycerol, DMSO, ethanol), high enzyme to DNA ratios, presence of divalent cations other than Mg^{++} (eg: Mn^{++}) and suboptimal storage conditions for the enzyme. Restriction enzymes are generally supplied with preformulated buffers and explicit directions for use.

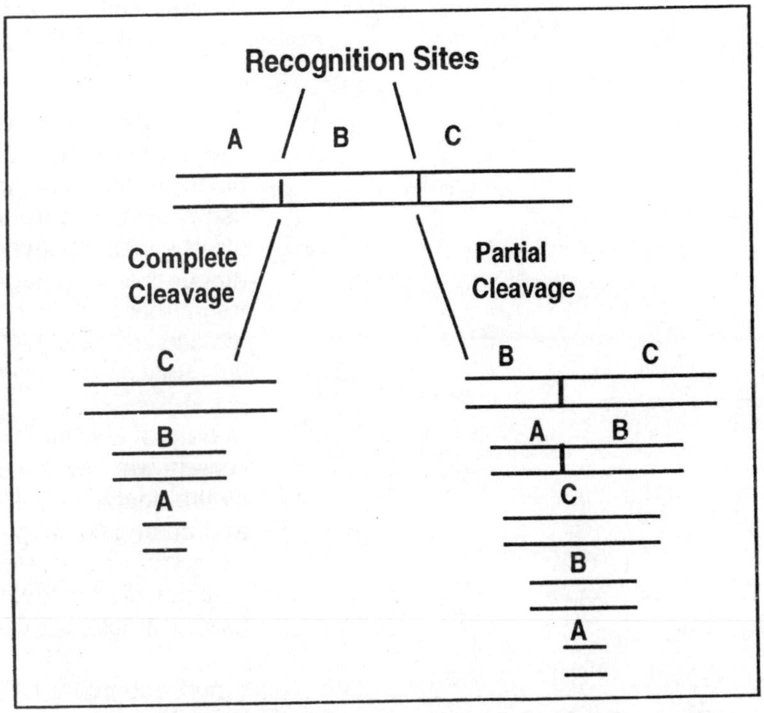

Figure 2-A diagrammatic comparison of partial and complete DNA cleavage by a single restriction endonuclease. A, B, and C fragments are produced by complete cleavage.

Introduction, continued

Some Considerations for this Laboratory Exercise

The key objective of this experiment is to gain an understanding of the restriction modification systems present in bacteria by cleaving or methylating different DNA molecules *in vitro*.

Specifically, in parallel manipulations, you will subject both plasmid and lambda DNA to digestion by each of the following restriction endonucleases: *Bgl* I, *Eco* RI, and *Hind* III. In addition, you will examine the putative protective affect of DNA methylation prior to treatment with the cognate restriction endonuclease (*Eco* RI).

For comparison, you will cut *E. coli* RT 13 chromosomal DNA with *Eco* RI and *Hind* III. Uncut controls for plasmid, lambda and chromosomal DNAs will be prepared.

Each of these preparations will be analyzed by agarose gel electrophoresis. The supercoiled plasmid DNA used as the main enzyme substrate contains approximately 4300 base pairs and has one recognition site for *Bgl* I and two for *Eco* RI and *Hind* III.

Lambda DNA is isolated as a linear molecule from the *E. coli* bacteriophage Lambda. It contains approximately 49,000 base pairs and has 5 recognition sites for *Eco* RI, 7 for *Hind* III and 29 for *Bgl* I.

The smallest of the fragments from the *Hind* III digest (125 base pairs) and several of the smaller fragments for the *Bgl* I digest will not be visible after agarose gel electrophoresis. Smaller fragments can run off the gel ahead of the marker dye. Consider also that a stoichiometric cleavage of a pure DNA sample results in equimolar amounts of each fragment. Since there is less mass in the smaller fragments, and since staining intensity is directly proportional to the molecular mass in the band, the faster bands will stain less intensely and may prove undetectable.

In nature, circular DNAs such as plasmids generally are "supercoiled", with twists either in addition to or less than those imposed by the double helix itself. Supercoiled DNA has a more compact and entangled shape (like a twisted rubber band) than its corresponding non-supercoiled forms. This helps it fit into the limited confines of a bacterial cell or eukaryotic organelle without the need for the structural molecules found in eukaryotic chromosomes.

When supercoiled DNA is nicked (a phosphate-sugar bond is broken anywhere in the molecule), it untwines to form a circle. A single cleavage of a supercoiled molecule by a restriction endonuclease results in a linear molecule.

Introduction, continued

During replication, two or more plasmids can form chains of interlocking rings called catenanes. Catenanes may be dimers (two rings), trimers (three rings) and so on.

Under the electrophoretic conditions used in this laboratory, for a given plasmid, the following relative electrophoretic mobilities apply:

faster ... *slower*

supercoiled > linear > nicked circular > dimer > trimer

The chromosomal DNA used in this lab was isolated from *E. coli* RY 13 (possessing the *Eco* RI restriction -modification enzymes) and contains about 3,000,000 base pairs. During isolation, random shearing of large DNAs such as this reduce the average fragment length to about 100,000 base pairs. You will compare the effect treatment of this DNA with R·*Eco* RI and R·*Hind* III.

Safety Guidelines

Safety glasses and gloves should be worn when performing this experimental module. Biological and other material should be disposed of in specially labeled containers. Hot agarose may splatter and can inflict severe burns. No food or drink should be used in the laboratory.

Materials

This module is based upon EDVOTEK experiment #205, Restriction-Modification of DNA. Experiment #205 contains one tube each of the following:

		Storage
A.	Restriction endonucleases	
	Bgl I	Freezer
	Hind III	Freezer
	Eco RI	Freezer
	Eco RI	Freezer
	Eco RI-Methylase	Freezer
B.	SAM	Freezer
C.	Plasmid DNA	Freezer
D.	Lambda DNA	Freezer
E.	*E. coli* RY 13 chromosomal DNA	Refrigerator
F.	Restriction enzyme reaction buffer	Freezer
G.	Methylase reaction buffer	Freezer
H.	Dilution buffer for methylase assay	Freezer

Concentrated gel loading solution
Practice gel loading solution
Agarose-nucleic acid grade
Electrophoresis buffer
Methylene Blue Plus™ stain
0.5 ml pipets
Transfer pipets
Microtest tubes with attached caps

Requirements

- Capillary pipets or automatic micropipet capable of delivering 50 µl
- Horizontal agarose gel electrophoresis apparatus with D.C. power supply
- 37°C water bath
- DNA visualization system (illuminator)
- Distilled or deionized water
- Insulated glove for handling hot agarose containers

Pre-lab Preparation

1. General Preparation

   a. Thaw all reaction buffers, DNAs, AdoMet and put them on ice. Tap the tubes with your fingers or on a table to get all the sample back to the bottom of the tube. Put on ice.

   b. Label 13 microtest tubes 1-13; Label 12 microtest tubes 2R-5R, 7R-10R, 12R, 13R, 5M, and 10M. Cap and put them on ice.

   Do not cross-contaminate enzyme and DNA stocks by using the same pipet.

2. Preparation of DNAs:

   The DNAs can be prepared several hours before the lab. Aliquot DNAs before you prepare the enzymes. DNA aliquots may be stored at room temperature on the day of the lab.

   a. Transfer 50 ml of **G** into each of tubes **1-5**. Cap.

   b. Using a fresh pipet tip, transfer 50 ml of **H** into each of tubes **6-10**. Cap.

   c. Using a fresh pipet tip, transfer 50 ml of **I** into each of tubes **11-13**. Cap.

   Many enzymes are labile. Keep them cold and minimize handling. Do not dilute enzymes into reaction buffers more than 30 min. before the lab. Once diluted, the enzymes must be used; they can no longer be stored. Remove enzyme stocks from the freezer (will still be liquid) and put them directly on ice shortly before they are required for pre-lab preparations.

3. Preparation of *Eco* RI methylase:

   a. Using a plastic transfer pipet, transfer all the ice cold buffer in tube **K** to tube **E**. Put tube **E** back on ice.

   b. Using a fresh plastic transfer pipet, transfer all the ice cold AdoMet in tube **F** to tube **E**. Cap tube **E**. Mix by tapping. Put back on ice.

   c. Using a fresh pipet tip, transfer 50 ml of enzyme solution in tube **E** into ice cold tubes **5M** and **10M**. Cap them and put back on ice. Dispose of used pipet tips.

Pre-lab Preparation, continued

4. Preparation of Restriction Enzymes:

a. With a fresh 0.5 ml pipet, add 0.15 ml of ice cold buffer in tube **J** to tube **A**. Cap tube **A**. Mix by tapping and put back on ice. Dispose of the pipet.

Note: There should not be any dense layer of enzyme solution left at the bottom of the stock tubes if properly mixed after dilution.

b. Transfer 50 ml from tube **A** to ice cold tubes **2R** and **7R**. Cap the tubes and put them back on ice. Dispose of the used tip.

c. With a fresh 0.5 ml pipet, add 0.2 ml of ice cold buffer in tube **J** to tube **B**. Cap tube **B**. Mix by tapping and put back on ice. Dispose of the pipet.

d. Transfer 50 ml from tube **B** to ice cold tubes **3R**, **8R** and **12R**. Cap and put back on ice. Dispose of the used tip.

e. Using a fresh 0.5 ml pipet, add 0.2 ml of ice cold buffer in tube **J** to tube **C**. Cap tube **C**. Mix by tapping and put back on ice. Dispose of the pipet.

f. Transfer 50 ml from tube **C** to ice cold tubes **4R**, **9R** and **13R** with pipet. Cap and put back on ice. Dispose of the used tip.

g. Using a fresh plastic transfer pipet, transfer all of the ice cold buffer in tube **L** to tube **D**. Cap tube **D**. Mix by tapping and put back on ice.

h. Transfer 50 ml from tube **D** into ice cold tubes **5R** and **10R**. Keep these tubes on ice until they are needed for a later incubation.

Method

General Instructions

1. Each complete experiment involves the correct mixing and electrophoresis of 13 tubes of reagents manipulated in parallel. Tubes 1-5 involve plasmid DNA, tubes 6-10 involve lambda DNA and the remaining three involve *E. coli* RY 13 chromosomal DNA. In an attempt to combine clarity and brevity in the limited space available on the surface of small tubes, the tube numbers often are combined with suffixes indicating their respective treatments. Tubes with both number and letter designations (R = restriction enzyme, M = *Eco* RI methylase) contain enzymes, reaction buffer and $MgCl_2$ (for the restriction enzymes) or SAM (for the methylase). Mixing the DNA with the enzyme solutions will start the reactions.

2. Use a fresh transfer pipet to mix the DNA with water or enzymes. Use only one transfer pipet per set of reactions to avoid cross-contamination. Unless otherwise indicated, transfer the entire contents from the tube containing DNA to the corresponding tube containing enzyme. Mix by tapping the tube after transfer. Tubes can be incubated in a 37°C water bath by suspending them in styrofoam floats.

3. Save the transfer pipets for the addition of concentrated gel loading solution and the loading of samples for electrophoresis. Note that gel loading solution includes protein denaturants that stop the enzyme reactions. Never place a pipet that has been in gel loading solution into a tube containing enzymes until you are ready to terminate the reaction.

4. Agarose gel 0.8% is prepared according to the general instruction outlined in Module 1, Unit 1.

Method, continued

Mixing Directions

Add Contents of	to Tube		Incubation 37°C, for	Well Contents
50 µl H_2O (distilled)	1	Cap.	30 min.	Plasmid DNA, Uncut
2	2R	Cap, mix.	30 min.	Plasmid DNA + R·*Bgl* I
3	3R	Cap, mix.	30 min.	Plasmid DNA + R·*Hind* III
4	4R	Cap, mix.	30 min.	Plasmid DNA + R·*Eco* RI
5	5M	Cap, mix.	20 min.	Plasmid DNA + M·*Bgl* I
5M	5R	Cap, mix.	20 min.	+ R·*Bgl* I
50 µl H_2O (distilled)	6	Cap.	30 min.	Lambda DNA, Uncut
7	7R	Cap, mix.	30 min.	Lambda DNA + R·*Bgl* I
8	8R*	Cap, mix.	30 min.	Lambda DNA + R·*Hind* III
9	9R	Cap, mix.	30 min.	Lambda DNA + R·*Eco* RI
10	10M	Cap, mix.	20 min.	Lambda DNA + M·*Bgl* I
10M	10R	Cap, mix.	20 min.	+ R·*Bgl* I
50 µl H_2O (distilled)	6	Cap.	30 min.	*E. coli* RY 13 chrom., uncut
12	12R	Cap, mix.	45 min.	*E. coli* RY 13 + *Hind* III
13	13R	Cap, mix.	45 min.	*E. coli* RY 13 + *Eco* RI

After all incubations are done, add 1 hanging drop (10 µl) of gel loading solution (N) to each tube. Cap and mix. These tubes may be stored in the freezer until ready for electrophoresis. Each tube should contain enough sample for two (2) electrophoresis runs.

*NOTE: Just before loading tube 8R, heat it to 65°C for 2 minutes, then allow to cool.

Results

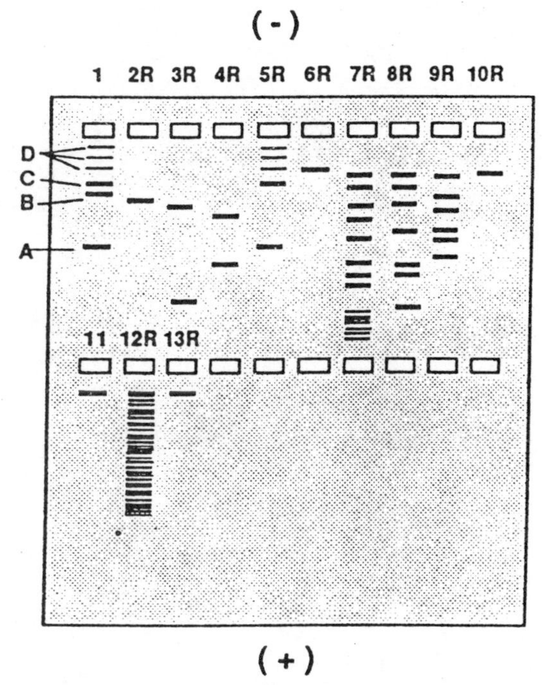

Sample 1 and 5R

A. Supercoiled

B. Nicked (may not be visible with methylene blue)

C. Dimer

D. Higher catenanes (may not be visible with methylene blue)

UNIT 2: RESTRICTION-MODIFICATION SYSTEMS

MODULE 8: CONSTRUCTION AND MAPPING OF RECOMBINANT PLASMID

Harley Mortensen

* Introduction
* Experimental Outline
* Safety Guidelines
* Materials
* Method
* Results

Introduction

The objective of this module is to insert a "foreign" gene into a plasmid vector and to demonstrate the gene transfer by screening resulting recombinants. Cloning vectors such as pTZ18U (United States Biochemical) contain a polycloning site (PCS), also referred to as a multiple cloning site (MCS) into which foreign DNA may be inserted.

MCS: AAGCTTGCATGCCTGCAGGTCGACTCTAGAGGATCCCCGGGTACCGAGCTCGAATTC

Hind III Sph I Pst I Sal I Acc I Hinc II Xba I Bam HI Sma I Xma I Kpn I Sac I Eco RI

Multiple Cloning Site

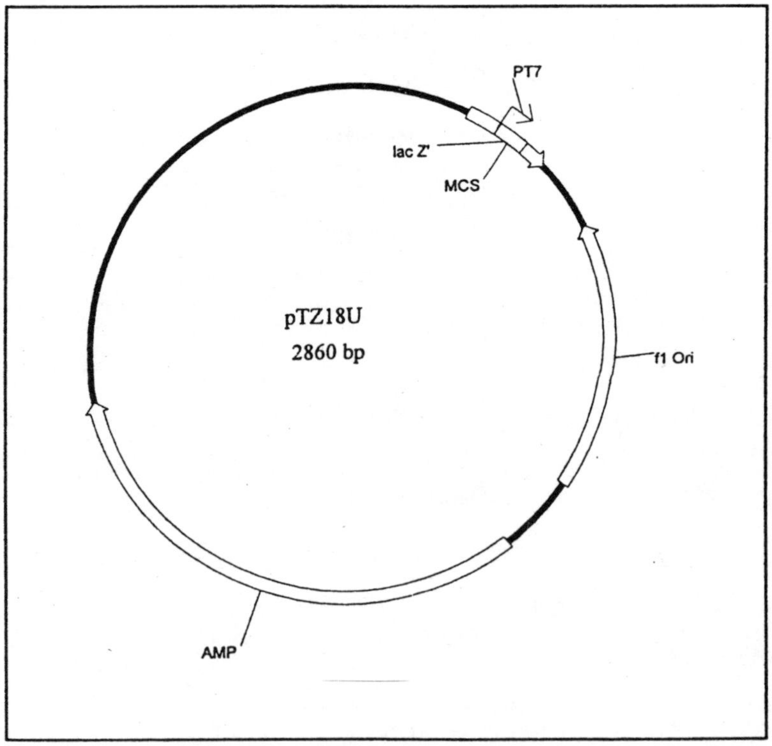

Figure 1. pTZ18U Restriction Map

Introduction, continued

This region consists of several restriction sites that do not appear elsewhere on the plasmid and are used separately or in combination to open the circular plasmid. If only one site is cut and, assuming the insert DNA has been cut with the same enzyme so that ends are compatible, there are two orientations of the insert that will occur. Figure 2 illustrates the situation using *Eco* RI to cut both the "foreign DNA" to be inserted and the plasmid. Unless the insert DNA carries its own promoter, typically only one of these orientations will contain the gene in an "expressible" orientation.

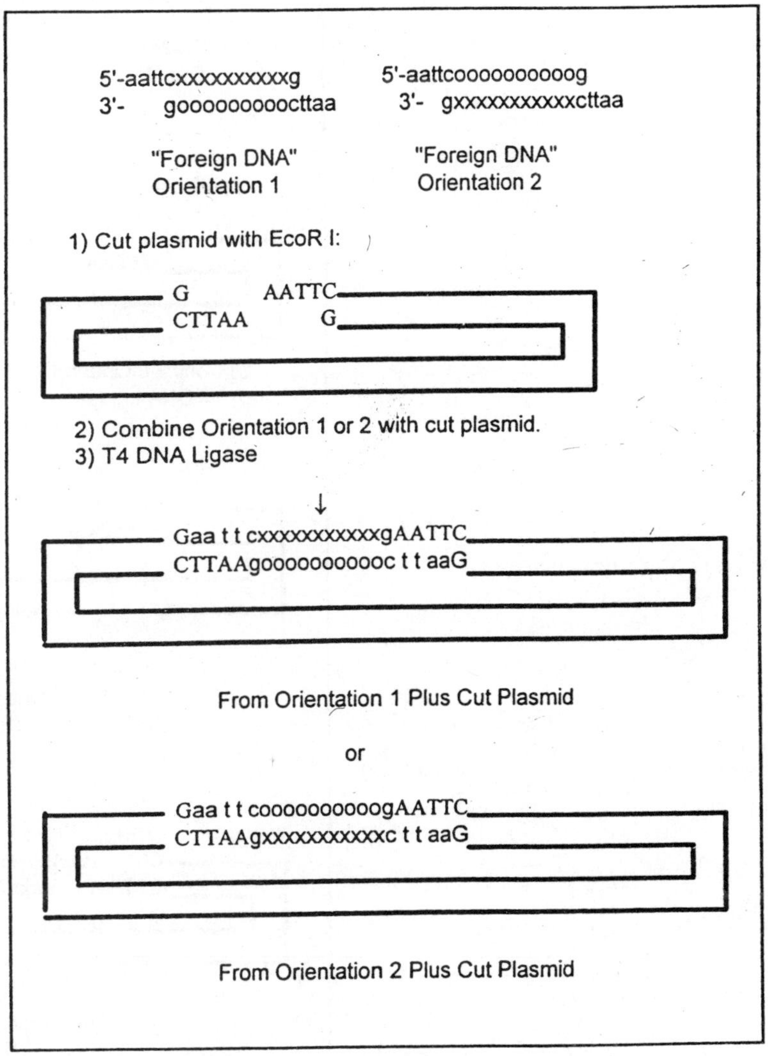

Figure 2. Cloning of "Foreign DNA" into Plasmid With All Ends (From *Eco* RI) Compatible Allows Two Insert Orientations.

Introduction, continued

If the plasmid is cut with two different restriction endonucleases, "forced cloning", in a directional sense, occurs such that the insert DNA can be ligated into the vector in only one orientation provided the insert DNA has been cut with the same two enzymes. This is shown in Figure 3 in which *Eco* RI and *Hind* III are used to cut both the 'Foreign DNA" and the plasmid.

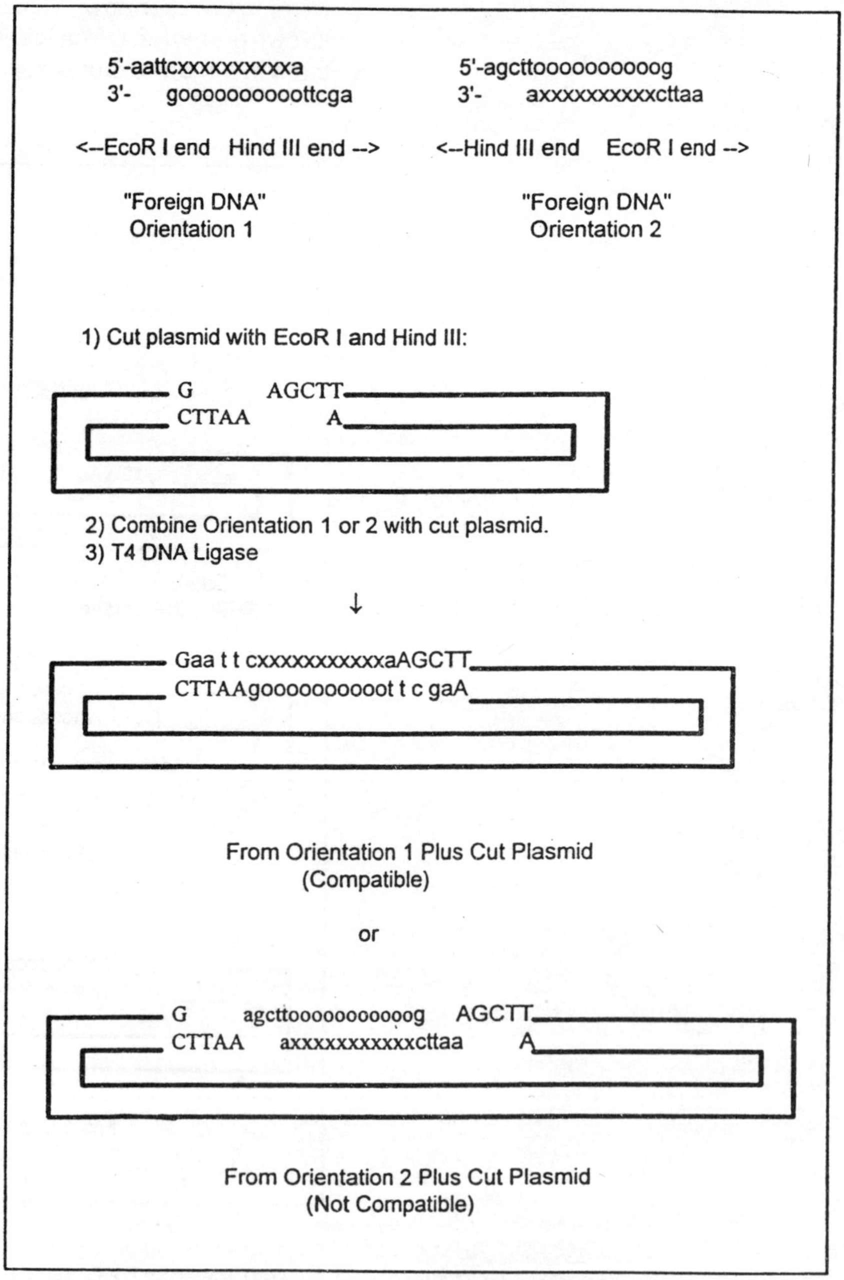

Figure 3. Forced Cloning of "Foreign DNA" into plasmid with *Eco* RI ends and *Hind* III ends allows only one insert orientation.

Introduction, continued

This exercise utilizes a 1300 bp kanamycin gene cartridge containing *Eco* RI ends. pTZ18U will be cut with *Eco* RI to generate cohesive ends identical to the kanamycin gene cartridge. If the two kinds of DNA are mixed, two orientations are possible which is less efficient than forced cloning. However, the screening procedure is powerful enough to detect the desired functional clones from a heterogeneous population of products. The compatible cohesive ends which occasionally anneal to form nicks can be joined by catalytic action of T4 DNA ligase. This is then used to transform *E. coli* strain DH11S which is screened for inserts by so-called "blue-white" screening and by resistance to kanamycin and ampicillin.

The lac Z' sequence is present in many of the vectors designed for recombinant DNA work. It codes for only the amino terminal amino acids of β-galactosidase. Thus, the expression of lac Z' in a plasmid will not lead to a functional enzyme. However, if a plasmid with lac Z' is within a host of genotype Δ lac/F' lacZΔ M15, lacI^q, the system can be used to screen for inserts cloned into the MCS. Hosts of this genotype also cannot generate a functional β-galactosidase but the combination of host and plasmid can do so by virtue of "complementation" between the plasmid lac Z' and host lac ZΔM15. Lac Z' of the plasmid leads to the amino terminal residues (alpha fragment) while lac Z Δ M15 of the host provides the larger carboxyl terminal residues (omega fragment). The two fragments together constitute a functional β-galactosidase. This is what is meant by the term "alpha complementation". The system can be regulated due to the presence of the constitutive mutation, lac I^q, which overproduces the lac repressor, resulting in the system being in the repressed state. The system is induced by being exposed to the synthetic inducer, isopropyl-β-thiogalactoside (IPTG). If the artificial substrate, X-gal (5-bromo-4-chloro-3-indolyl-β-D-galactoside), is present, it will be hydrolyzed by β-galactosidase upon induction to a blue dye. If DNA is inserted within the MCS, the lac Z' sequence is disrupted. In this state, the resulting "alpha" fragment amino acid sequence is no longer capable of complementation. Thus, host cells harboring plasmids with no inserts will undergo alpha complementation and produce blue colonies while those with inserts will not alpha complement and will produce white colonies since the lacZ gene is interrupted.

The screening procedure to detect recombinant transformants is based on having the Amp gene and the Kan gene present in the recombinant plasmid. In this exercise, the correct transformants will carry the kan gene as an insert within the MCS which is itself located within the lac Z' sequence. This situation allows for the application of "blue-white" screening, since the host cell, DH11S, is of the requisite genotype. The recombinant transformants are also amp^r and kan^r.

Experimental Outline

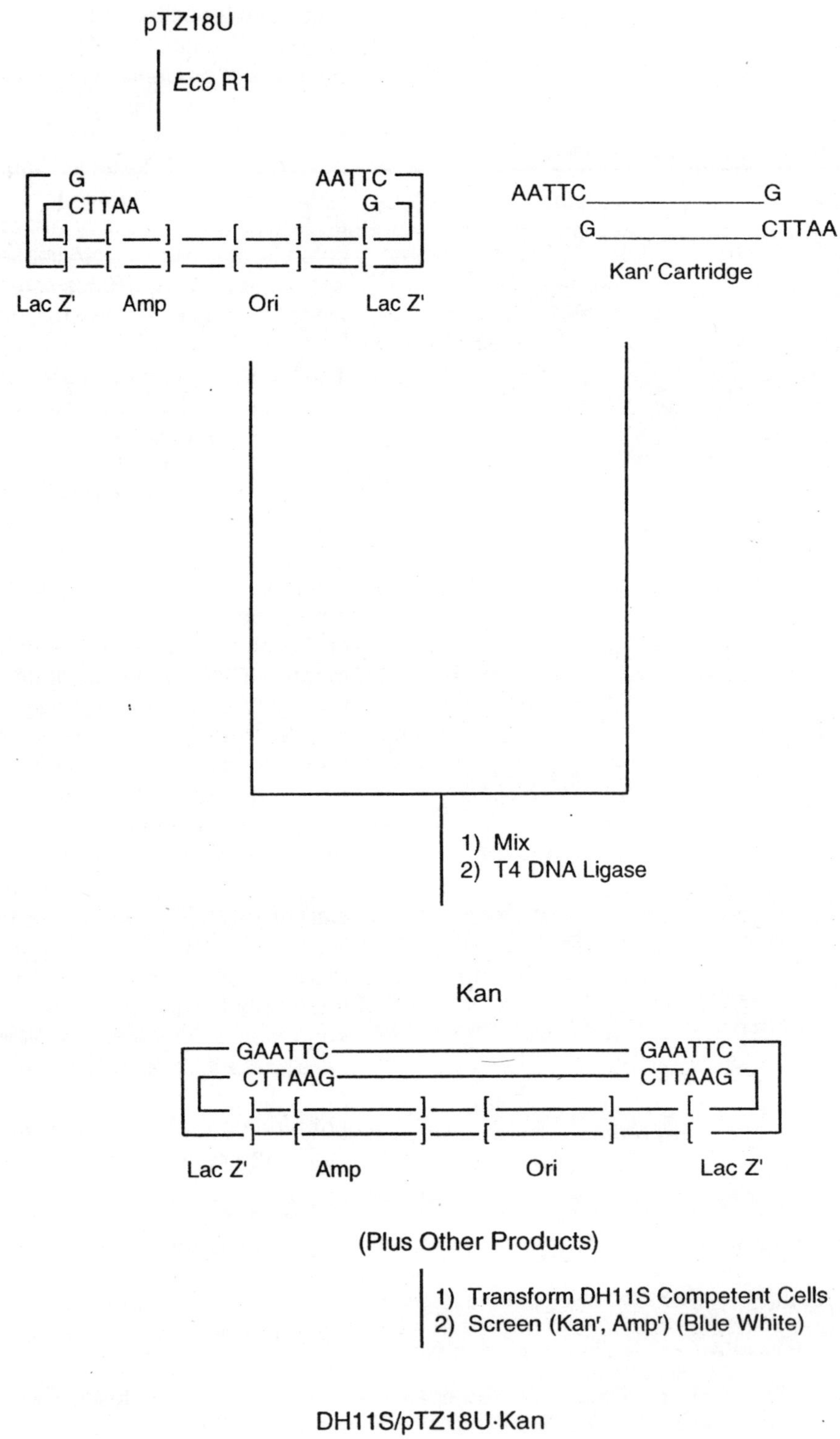

Safety Guidelines

Use gloves when handling ethidium bromide. Ethidium Bromide is a potent mutagen. Use gloves and goggles when handling phenol. Phenol causes burns when in contact with tissue. Use goggles or other eye protection when viewing gels through the transilluminator. Ultraviolet radiation from the transilluminator can cause tissue damage.

Materials

Step 1 - Vector Preparation

pTZ18U (United States Biochemical)
Eco RI
Hind III
Lambda DNA (0.5 µg/µl)
Tracking Dye
Ethidium Bromide (10 µg/ml)
3 M Sodium Acetate
95% Ethanol
1.5 ml Microfuge Tubes
Sterile ddH_2O
Gloves
Goggles
Pipets
Electrophoresis Cell
Power Supply
Microfuge
37°C Water Bath
Transilluminator

Materials, continued

Step 2 - Ligation

pTZ18U *Eco* RI Digest
Kanamycin Cartridge (Kan^{R}GenBlock)
(Pharmacia LKB Biotechnology)
5X Ligation Buffer
T4 DNA Ligase
1.5 ml Microfuge Tubes
Sterile ddH_2O
3 M Sodium Acetate
95% Ethanol
Agarose
Ethidium Bromide
Polaroid Film 667
DH11S Competent Cells
SOC Medium
Pipets
12°C Water Bath
Microfuge
Transilluminator
Camera
Incubator

Step 3 - Screening
(Blue-White, Amp^r, Kan^r)

Nutrient Agar
Ampicillin Stock (50 mg/ml in water)
Kanamycin Sulfate Stock (50 mg/ml in water)
X-Gal 2% in H_2O
IPTG 2% in Dimethyl Formamide
Sterile Saline (0.85%)
Glass Spreaders
Incubator

Materials, continued

Step 4: Restriction Digestion and Gel Analysis of Transformants

DH11S/pTZ18U·Kan
LB Broth
Kanamycin (50 mg/ml)
TELT Solution
  50 mM Tris-cl, pH 8
  62.5 mM Na_2EDTA
  2.5 M LiCl
  4% (vol/vol) Triton X-100
1:1 (vol/vol) Phenol:Chloroform
RNase A
TE Buffer
Hind III
Eco RI
Cla I
Dra III
Lambda DNA (0.5 μg/μl)
Tracking Dye
80% Glycerol
Ethidium Bromide
3 M Sodium Acetate
95% Ethanol
1.5 ml Microfuge Tubes
Sterile ddH_2O
Gloves
Goggles
Pipets
Electrophoresis Cell
Power Supply
37°C Water Bath
Microfuge
Transilluminator
Goggles/Face Shield

Method

Step 1: Vector Preparation

In this procedure, the plasmid pTZ18U is cut with *Eco* RI to linearize the plasmid and generate *Eco* RI "sticky ends" within the MCS. The pTZ18U can either be purchased or isolated from stocks such as DH11S/pTZ18U, if available. If the vector has been isolated from stocks, it should be free from RNA contamination by treatment with RNase A.

RNase A Digestion: Prepare a solution containing 10 mg/ml of RNase A. Heat in a boiling water bath for ten minutes to inactivate any DNase activity. Cool slowly. Store at -20°C. Add to a plasmid preparation to a concentration of 10-20 µg/ml and incubate for 30 minutes. Remove protein by phenol extraction followed by ethanol precipitation. Centrifuge and dissolve the pellet in TE buffer or sterile ddH_20.

1. Set up restriction digestions and controls as follows using three sterile 1.5 ml microfuge tubes:

	TUBE 1	TUBE 2	TUBE 3
pTZ18U in ddH_20	20 µl	-	5 µl
10x core buffer	3 µl	1 µl	-
Sterile water ddH_20	6 µl	7 µl	5 µl
Eco RI	1 µl	-	-
Hind III	-	1 µl	-
Lambda DNA (.5 µg/µl)	-	1 µl	-

   Tube 2 is a size standard for estimating sizes of restriction fragments generated by the digestion.

   Enzyme buffer for *Eco* RI: 50 mM Tris HCl, pH8.0; 10 mM $MgCl_2$; 100 mM NaCl

   Enzyme buffer for *Hind* III: 50 mM Tris HCl, pH 8.0; 10 mM $MgCl_2$; 50 mM NaCl

   Centrifuge briefly in a microfuge to mix.

2. Incubate at 37°C for 30-45 minutes.

3. After the incubation period, transfer 5 µl from tube 1 and tube 2 to separate fresh tubes, add 2 µl tracking dye to each and load in separate lanes of a gel. Load the contents of tube 3 to a third gel lane after adding 2 µl tracking dye to the tube.

Method, continued

Step 1: Vector Preparation, continued

4. Perform electrophoresis as described in Unit 1, Module 1.

5. To the remainder of tube 1, add 2.5 µl of 3 M sodium acetate and 50 µl ethanol to precipitate DNA. Place at -20°C for at 20 mins.

6. Centrifuge the tube at maximum speed in a microfuge for 5 minutes to pellet the DNA.

7. Decant the supernatant and dry the pellets.

8. Dissolve in 20 µl of sterile ddH_20 and store at 4°C.

Step 2: Ligation

1. To a sterile 1.5 ml microfuge tube, add 10 µl each of *Eco* RI digest of pTZ18U and 5 µl of the Kan cartridge DNA, 6 µl of 5X ligation buffer, and 8 µl sterile ddH_2O. Remove 10 µl to be used as a control. To the remainder, add 1 µl T4 DNA ligase and incubate at 12°C overnight.

2. Check the ligation by running a gel using 5 µl of the ligation reaction and 5 µl of unligated control mixture. Add 2 µl 80% glycerol to each for loading in the gel wells. Do not use tracking dye because the dye sometimes obscures the kanamycin DNA fragment. A separate lane loaded with *Hind* III cut Lambda DNA with tracking dye can be used to follow the course of the run. Alternatively, simply run the gel for one hour at 100 V.

3. Heat the remainder of the ligation mixture in a 75°C water bath for 5 minutes to denature proteins.

4. Add 2 µl 3 M sodium acetate and 40 µl 95% ETOH, chill 20 minutes at -20°C and centrifuge to pellet the DNA.

5. Pour off the supernatant and dry the pellet.

6. Dissolve the pellet in 50 µl of TE buffer and store in the refrigerator.

7. Do a quantitative estimation of the DNA concentration at 260 nm as described below or by using a series of DNA standards of known concentration stained with ethidium bromide on an agarose gel.

Method, continued

Step 2: Ligation, continued

Quantitative Estimation of DNA

DNA concentration can be determined by measuring absorbance of a solution at 260 nm. Use the relation: 1 A_{260} unit = 50 μg/ml. Small volumes such as are used in recombinant DNA work require a 5 carat microcell (Beckman). An alternative procedure calls for casting an agarose gel and forming a series of wells in the surface. Dilutions of DNA from a stock of known concentration are made, small volumes are added to the wells and allowed to penetrate the gel. An equal volume of unknown is added to a well. The gel is stained with ethidium bromide, placed on a transilluminator and an estimate of the unknown concentration is made by comparison with the knowns. The gel can also be photographed and the photograph used for making the comparison.

The preparation from above will be used to transform host cells which can be screened for the proper recombinants. The transformation procedure is described elsewhere in this manual. If competent cells of DH11S were prepared earlier, they should be used for this exercise. Otherwise, cells will have to be made competent (Unit 3, Module 12) or purchased.

The transformed cells can be used to demonstrate blue-white screening. Also, transformants will be screened for antibiotic resistance to ampicillin and kanamycin.

Step 3: Screening (Blue-White, Amp^r, Kan^r)

The following agar plates are to be prepared in advance.

1 LB
2 LB/ampicillin (50 μg/ml)
2 LB/kanamycin (50 μg/ml)

Pipet 80 μl of X-Gal and 20 μl IPTG on each of the two plates containing ampicillin. Spread evenly over the surface as when spreading cells. Allow to dry for several hours.

1. After the one-hour incubation period following transformation, using sterile 0.85% saline as diluent, prepare 1 ml each of 10^{-1}, 10^{-2}, 10^{-6} dilutions of the cell suspension. (For each ten-fold dilution, add 100 μl to 900 μl diluent.)

Method, continued

Step 3: Screening (Blue-White, Amp^r, Kan^r) , continued

2. Spread 100 µl of the 10^{-6} dilution on the LB plate. This is a control plate to determine the number of viable cells in the suspension.

3. Spread 100 µl of 10^{-1} and 10^{-2} dilutions on LB/ampicillin plates. This allows a determination of the number of transformants.

4. Spread 100 µl of 10^{-1} dilution and 100 µl of undiluted cell suspension on the RA/Kan plates This allows a determination of the number of transformants due to plasmids which contain both the kanamycin gene and the ampicillin gene.

5. Incubate the plates at 37°C until cell growth occurs.

6. Count all colonies on all plates. Record the number of white versus the number of blue colonies on the plates containing IPTG and X-Gal.

7. Calculate the ratio of cells transformed by recombinant plasmids to total transformants.

Step 4 - Restriction Digestion and Gel Analysis of Transformants

In this part of the exercise, a few colonies will be selected from the LB/Kan plates. Plasmid will be isolated from each and subjected to restriction digestion by *Eco* RI. Digested and undigested plasmids will be compared by gel electrophoresis.

Select six colonies from the LB/Kanamycin plates. Select a blue colony from a LB/Amp plate. Isolate plasmid according to the procedure in Module 2: Plasmid Isolation.

1. In a 1.5 ml microfuge tube, set up a restriction digestion with *Eco* RI and plasmid preparation isolated from a LB/kanamycin colony expected to contain recombinant plasmid.

2. Set up a second tube as above but use a blue colony plasmid.

Method, continued

Step 4: Restriction Digestion and Gel Analysis of Transformants, continued

3. Set up five tubes using a different recombinant plasmid preparation in each.

Plasmid solution	10 µl
10x core buffer	2 µl
ddH_2O	6 µl
Cla I	1 µl
Dra III	1 µl

4. Set up a tube to digest lambda to be used as a size marker.

Lambda DNA	3 µl
10x core buffer	1 µl
ddH_2O	5 µl
Hind III	1 µl

5. Incubate all tubes at 37 °C for 30-45 minutes.

6. Except for the *Hind* III digest, add 2 µl 80% glycerol to each digest. Do not use tracking dye in digest as it will tend to obscure the kanamycin gene and other small fragments after gel electrophoresis. Add 2 µl of tracking dye to the *Hind* III digest.

7. Load 5-10 µl from each of the eight tubes in separate lanes with the *Hind* III digest near the center to facilitate size determination after electrophoresis.

8. Electrophorese at 100 V until the tracking dye in the *Hind* III lane is approximately one inch from the end of the gel.

9. Stain in ethidium bromide for ten minutes, destain ten minutes and photograph on the transilluminator.

Results

Using the *Hind* III pattern as a size standard, verify the 1300 bp kanamycin fragment present in the *Eco* RI digest of the recombinant plasmid preparation. Compare with the blue (non-recombinant) colony digest.

Determine fragment sizes of the *Cla*I/*Dra* III double digests. Is there evidence for more than one orientation of the kanamycin insert?

UNIT 2: RESTRICTION-MODIFICATION SYSTEMS

MODULE 9: DNA FINGERPRINTING

Jack G. Chirikjian

* Introduction
* Safety Guidelines
* Experimental Outline
* Materials
* Pre-lab Preparation
* Method
* Results

Introduction

DNA typing (also called DNA profile analysis or DNA fingerprinting) is a recently developed method that allows for the unambiguous identification of the source of unknown DNA samples. The method has become very important in forensic biochemical laboratories where it has been used to provide evidence in paternity and criminal cases. In contrast to the more conventional methodologies, such as blood typing, which can only exclude a suspect, DNA fingerprinting can provide positive identification with great accuracy.

DNA fingerprinting involves the electrophoretic analysis of DNA fragment sizes generated by restriction enzymes. Restriction enzymes are endonucleases which catalyze the cleavage of the phosphate bonds within both strands of DNA. They require Mg^{+2} for activity and generate a 5 prime (5') phosphate and a 3 prime (3') hydroxyl group at the point of cleavage. The distinguishing feature of restriction enzymes is that they only cut at very specific sequences of bases. Restriction enzymes are produced by many different species of bacteria (including blue-green algae). Over 1500 restriction enzymes have been discovered and catalogued.

Restriction Enzyme	Organism
Bgl I	*Bacillus globigii*
Bam HI	*Bacillus amyloliquefaciens* H
Eco RI	*Escherichia coli*, strain RY 13
Eco RII	*Escherichia coli*, strain R 245
Hae III	*Haemophilus aegyptius*
Hind III	*Haemophilus influenzae* R_d

Restriction enzymes are named according to the organism from which they are isolated. This is done by using the first letter of the genus followed by the first two letters of the species. Only certain strains or sub-strains of a particular species may produce restriction enzymes. The type of strain or substrain sometimes follows the species designation in the name. Finally, a Roman numeral is always used to designate one out of possibly several different restriction enzymes produced by the same organism or by different substrains of the same strain.

A restriction enzyme requires a specific double-stranded recognition sequence of nucleotide bases to cut DNA. Recognition sites are generally 4 to 8 base pairs in length. Cleavage occurs within or near the site. The cleavage positions are indicated by arrows. Recognition sites are frequently symmetrical, i.e., both DNA strands in the site have the same base sequence when read 5' to 3'. Such sequences are called palindromes. Consider the recognition site and cleavage pattern of *Eco* RI as an example.

```
     ↓
5'-GAATTC-3'        5'-G      AATTC-3'
3'-CTTAAG-5'        3'-CTTAA      G-5'
         ↑
```

As shown above, *Eco* RI causes staggered cleavage of its site. The resulting ends of the DNA fragments are called "sticky" or "cohesive" ends. This is because the single-stranded regions of the ends are complementary.

Introduction, continued

Some restriction enzymes, such as *Hae* III, introduce cuts that are opposite each other. This type of cleavage generates "blunt" ends.

```
        ↓
   5'-GGCC-3'        5'-GG      CC-3'
    3'-CCGG-5'       3'-CC      GG-5'
        ↑
```

The recognition sites of some restriction enzymes contain variable base positions. For example, *Ava* I recognizes:

```
     ↓
 5'-CPyCGPuG-3'     (Py = pyrimidine = C or T and
 3'-GPuGCPyC-5'      Pu = purine = G or A)
           ↑
```

Keep in mind that A pairs with T and G pairs with C. Consequently, there are four possible sequences *Ava* I recognizes. Recognition sites of this type are called degenerate.

There are some recognition sites that are divided by a certain number of totally variable bases. For example, *Bgl* I recognizes:

```
              ↓
 5'-GCCNNNNNGGC-3'        (N = A, G, C or T)
 3'-CGGNNNNNCCG-5'
         ↑
```

There are 625 possible sequences *Bgl* I can cleave. The only bases the enzyme truly "recognizes" are the six G-C base pairs at the ends, which forms a palindrome. In the case of *Bgl* I, these true recognition bases must always be separated by 5 base pairs of DNA, otherwise the enzyme cannot properly interact with the DNA and cleave it. Recognition sites like that of *Bgl* I are called hyphenated sites.

The size of the DNA fragments generated depends on the distance between the recognition sites. In general, the longer the DNA molecule, the greater the probability that a given recognition site will occur. The DNA of an average human chromosome is very large, containing over 100 million base pairs. A restriction enzyme having a 6-base pair recognition site, such as *Eco* RI, would be expected to cut human DNA into approximately 750,000 different fragments.

No two individuals have exactly the same pattern of restriction enzyme recognition sites. There are several reasons for this fact. A large number of alleles exist in the population. Alleles are alternate forms of a gene. Alleles result in alternative expressions of genetic traits which can be dominant or recessive. Chromosomes occur in matching pairs, one of maternal and the other of

Introduction, continued

paternal origin. The two copies of a gene (which can be alleles) at a given chromosomal locus, and which represent a composite of the parental genes, constitute an individual's unique genotype. It follows that alleles have differences in their base sequences which consequently creates differences in the distribution and frequencies of restriction enzyme recognition sites. Other differences in base sequences between individuals can occur because of mutations and deletions. Such changes can also create or eliminate a recognition site. The example in Figure 1 shows how a silent mutation can eliminate a recognition site but leave a protein product unchanged.

Eco RI site	
5'——ACG AAT TCC——3'	Coding DNA
H_2N——Thr - Asn - Ser——COOH	Protein
5' —— ACG AAC* TCC——3'	Codon changed by mutation
H_2N ——Thr - Asn - Ser——COOH	Protein

Figure 1

Individual variations in the distances between recognition sites in chromosomal DNA are often caused by intervening repetitive base sequences. Repetitious sequences constitute a large fraction of the mammalian genome and have no known genetic function. These sequences can occur between genes or are adjacent to them. They are also found within introns. Ten to fifteen percent of mammalian DNA consists of sets of repeated, short sequences of bases that are tandemly arranged in arrays. The length of these arrays (the amount of repeated sets) varies between individuals at different chromosomal loci.

TGTTTA|TGTTTA|TGTTTA|.........variable number

When these arrays are flanked by recognition sites, the length of the repeat will determine the size of the restriction enzyme fragment generated. There are several types of these short, repetitive sequences and they have been cloned and purified.

The variations in DNA sequences between individuals as determined by differences in restriction enzyme cleavage patterns are known as restriction fragment length polymorphisms (RFLPs). RFLPs are a manifestation of the unique molecular genetic profile, or "fingerprint", of an individual's DNA.

RFLP analysis of genomic DNA is facilitated by Southern blot analysis. After electrophoresis, the DNA fragments in the gel are denatured by soaking in an alkali solution. This causes double-stranded fragments to be converted into single-stranded form (no longer base-paired in a double helix). A replica of the electrophoretic pattern of DNA fragments in the gel is made by transferring (blotting) them to a sheet of nitrocellulose or nylon membrane. This is done by placing the membrane on the gel after electrophoresis

Introduction, continued

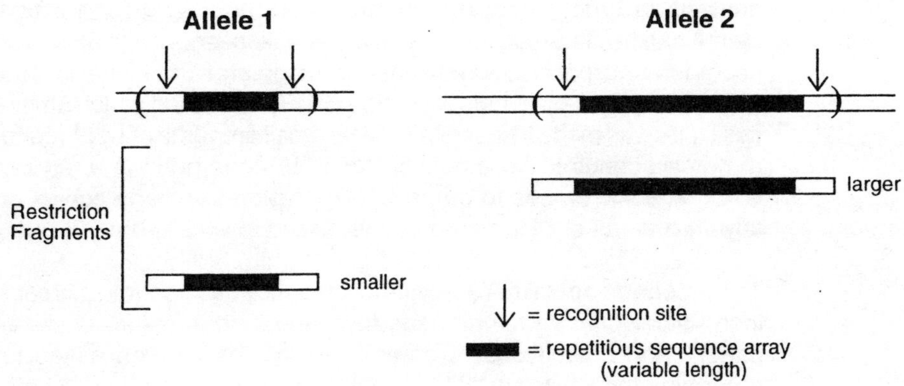

Figure 2

and transferring the fragments to the membrane by capillary action or suction by vacuum. The DNA, which is not visible, becomes permanently adsorbed to the membrane, which can be manipulated much more easily than gels.

Analysis of the blotted DNA is done by hybridization with a radioactive DNA probe. In forensic RFLP analysis, the probe is a DNA fragment that contains base sequences which are complementary to the variable arrays of tandemly repeated sequences found in the human chromosomes. Probes can be labeled with isotopic or non-isotopic reporter molecules that are used for detection. A solution containing the single-stranded form of the probe is incubated with the membrane containing the blotted, single-stranded (denatured) DNA fragments. Under the proper conditions, the probe will only base pair (hybridize) to those fragments containing the complementary repeated sequences. The membrane is then washed to remove excess probe. If the probe is isotopically labeled to the membrane, it is then placed on an x-ray film for several hours. This process is known as autoradiography. Only DNA fragments that have hybridized to the probe will reveal their positions on the film because the localized areas of radioactivity cause exposure. The hybridized fragments appear as discrete bands (fingerprint) on the film and are in the same relative positions as they were in the agarose gel after electrophoresis. Only specific DNA fragments of the hundreds of thousands of fragments present, will hybridize with the probe because of the selective nature of the hybridization process. Since autoradiography is an extremely sensitive technique, very small amounts of DNA samples are required.

In forensic cases, DNA samples can be extracted and purified from small specimens of skin, blood, semen, or hair roots collected at the crime scene. DNA that is suitable for analysis can even be obtained from dried stains of semen and blood. The RFLP analyses performed on these samples is then compared to those performed on samples obtained from the suspect. If the RFLP patterns match, although not absolute, it is beyond reasonable doubt that the suspect (or biological material from the suspect, such as blood) was at the crime scene. In practice, several different probes containing

Introduction, continued

different types of repetitious sequences are used in DNA profile analysis in order to satisfy certain statistical criteria for positive identification. The use of different restriction enzymes allow for accuracies in positive identifications of greater than one in 100 million. Polymerase Chain Reaction (PCR), a method which amplifies DNA, has made it possible for very small amounts of DNA found at crime scenes to be amplified for DNA fingerprinting analysis. Using specific probes to prime DNA polymerase, many copies of targeted areas of DNA can be synthesized *in vitro* and analyzed.

The presentation of DNA analysis as evidence has become increasingly significant in court cases involving murder, rape, physical battery and other types of crimes. Jurors have been asked to determine the validity of DNA evidence, which has resulted in both acquittals and convictions of suspected criminals. To ensure greater accuracy, scientists incorporate standardization procedures in DNA analysis. Routinely, Standard DNA Fragments are used to determine the exact size of individual DNA fragments in a DNA fingerprint. It is generally accepted that DNA fingerprints are identical only in the case of identical twins. Thus, defense and prosecuting attorneys debate the statistical probabilities of the DNA evidence as it applies to the general population. Although DNA analysis is not considered absolute, this technology has become an increasingly important part of our daily life.

Another application of DNA typing is paternity determination. A child's DNA is a composite of its parents' DNA. Therefore, comparison of DNA fragmentation patterns obtained from the mother and child will give a partial match. Bands in the child's fingerprint that are not present in the mother's must have been contributed by the father. In other words, because of allelic differences, not all bands present in the parents' fingerprint appear in the child's fingerprint. However, bands that do appear in the child's fingerprint must be found in either the father's or mother's fingerprint.

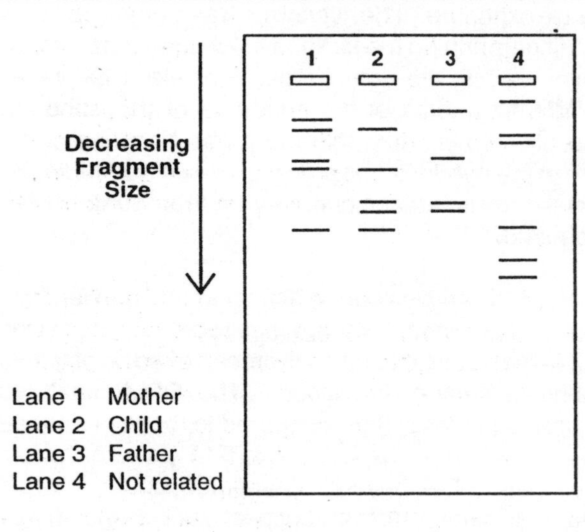

Figure 3

In DNA fingerprinting laboratories, the two most commonly used restriction enzymes are *Hae* III and *Hinf* I, which are 4-base and 5-base cutting enzymes.

Hae III

↓
5'-GGCC-3'
3'-CCGG-5'
↑

Hinf I

↓
5'-GANTC-3'
3'-CTNAG-5'
↑

Introduction, continued

This experiment simulates a hypothetical forensic case. In this module, DNAs from the hypothetical crime scene and suspects are cut by six-base cutting enzymes. The objective is to analyze the DNA fragmentation patterns compared with "crime scene" samples after agarose gel electrophoresis and determine if Suspect 1 or 2 was at the crime scene. DNA was obtained from the crime scene and two suspects. Each suspect's DNA will be cleaved with two restriction enzymes in separate reactions, and pairs of fragmentation patterns will serve as the individual fingerprint. The DNA fragmentation patterns are simple enough to analyze directly in the stained agarose gel, which eliminates the need for a Southern blot.

Safety Guidelines

This experiment is designed for staining of DNA with Methylene Blue Plus™ stain after electrophoresis. Methylene Blue Plus™ is formulated to provide optimal sensitivity for visualization. Enhanced visualization is obtained by using a Visible Light Gel Visualization system. As with any biological stain, care should be taken when handling solutions or gels containing methylene blue.

Gloves and goggles should be worn when handling methylene blue staining reagents, and worn routinely throughout the experiment as good laboratory practice.

Experimental Outline

DNA from two suspects are each cut with two restriction enzymes in separate reactions and compared to crime scene samples after agarose gel electrophoresis. This flow chart diagrams the procedure used for restriction enzyme digestion of DNA obtained from Suspect 1. DNA from Suspect 2 is digested in the same manner.

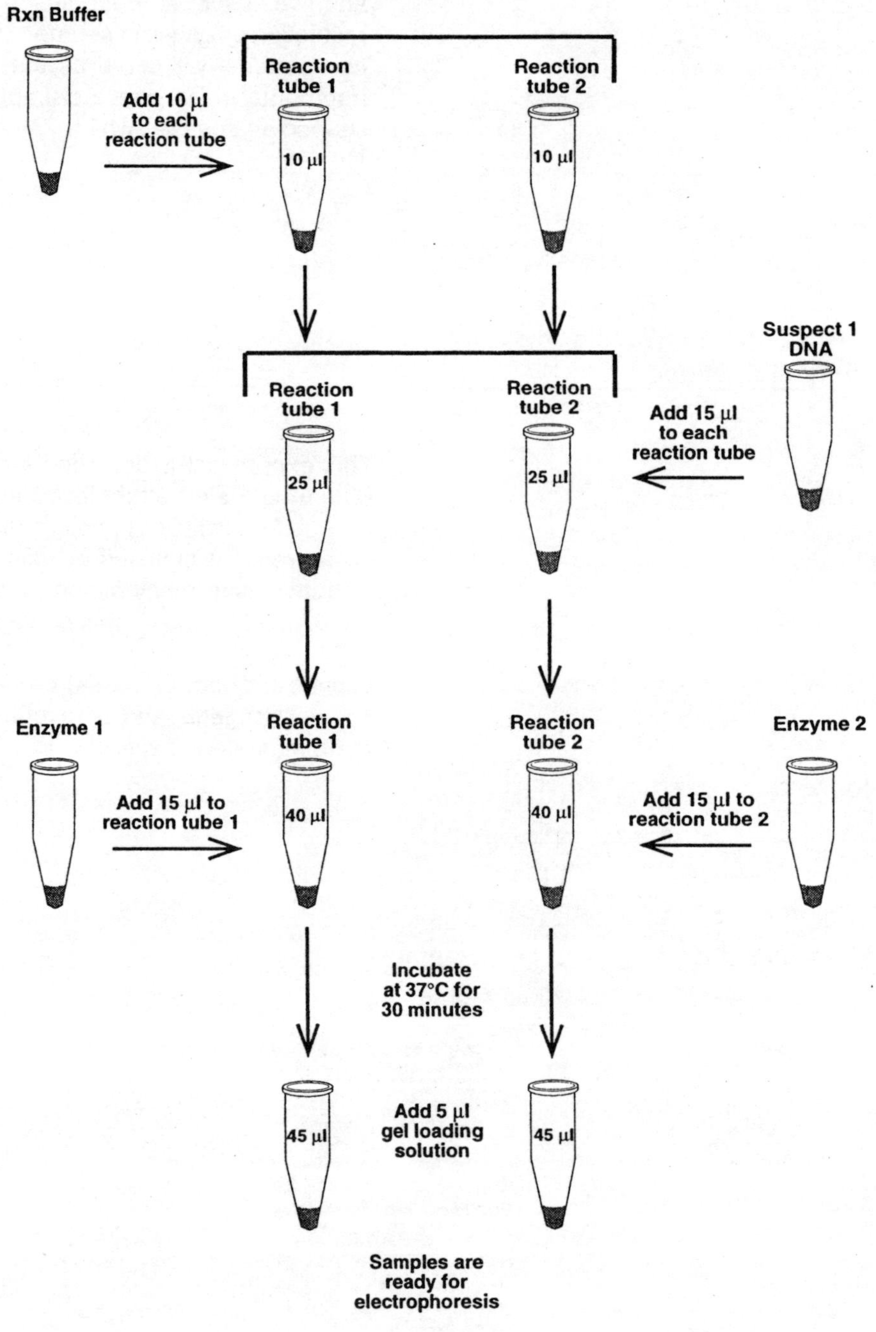

Materials

This module is based on EDVOTEK experiment #225. The components include the following:

Components A - H should be stored in a -20°C freezer.

A Crime scene DNA sample, pre-cut with Restriction Enzyme 1
B Crime scene DNA sample, pre-cut with Restriction Enzyme 2
(Samples A and B are ready for electrophoresis)
C Suspect #1 DNA sample
D Suspect #2 DNA sample
E Standard DNA Fragments
F Enzyme Dilution/Reaction Buffer
G Restriction Enzyme 1
H Restriction Enzyme 2

The following components can be stored at room temperature.

1 Tube 10x Gel Loading Solution
1 Tube Practice Gel Loading Solution
1 Bottle of UltraSpec-Agarose™ powder
1 Bottle of 50x concentrated electrophoresis buffer
1 Bottle of concentrated Methylene Blue Plus™ stain
3 1 ml pipet
1 100 ml plastic graduated cylinder
12 Transfer pipets
60 Microtest tubes with attached caps

UltraSpec-Agarose™ and Methylene Blue Plus™ are trademarks of EDVOTEK, Inc.

Requirements

- Horizontal gel electrophoresis apparatus
- D.C. power supply
- Automatic micropipets with tips
- Water bath (37°C)
- Recommended equipment:
  - DNA visualization system (white light)
  - Staining Tray and Net
- 250 ml flasks
- Pipet pump
- Hot gloves
- Marking pens
- Distilled or deionized water
- Ice

Pre-lab Preparation

Preparation of Biologicals and Reagents for Restriction Enzyme Digestion

1. Thaw the Enzyme Dilution/ Reaction Buffer (F) and all DNAs on ice. Tap the tubes with your fingers or on a table to get all the sample to the bottom of the tube.

2. Two tubes, components labeled A and B, contain the crime scene samples. These DNA samples have already been cut with restriction enzymes and are ready for electrophoresis. Sample A represents "crime scene" DNA cut with Restriction Enzyme 1. Sample B represents "crime scene" DNA cut with Restriction Enzyme 2.

   - Label 5 tubes "CS 1" for the crime scene sample #1 (A)
   - Label 5 tubes "CS 2" for the crime scene sample #2 (B)
   - Dispense 45 µl of each crime scene sample in the appropriate tubes for each of the five lab groups.

3. Component E contains Standard DNA fragments (markers).

   - Label five tubes "Markers".
   - Dispense 85 µl of Standard DNA fragments to each tube for each of the 5 groups.

4. Component F is the Enzyme Dilution/ Reaction buffer.

   - Label five tubes "Rxn Buffer".
   - Dispense 45 µl of Enzyme Dilution/ Reaction buffer to each tube for each of the 5 groups.

Preparation of Suspect DNA

5. Using an automatic micropipet, dispense the two Suspect DNAs (C, D) for each of the five lab groups.

   - For each of 5 groups, label a set of two tubes: "DNA 1", and "DNA 2".
   - Dispense 35 µl of each Suspect DNA to appropriate tubes.

Preparation of Restriction Enzymes

Prepare enzymes approximately 30 minutes before experiment. Enzyme stocks will still be liquid when removed from freezer storage. Place them on ice when ready to dilute the enzymes.

6. Dilute and Dispense Enzyme #1:

   - With a clean, dry 1 ml pipet, add 0.25 ml of **ice cold** Enzyme Dilution/Reaction Buffer (F) to Restriction Enzyme 1 (G).

Pre-lab Preparation, continued

- Mix by gently tapping until the denser layer of glycerol solution (in the enzyme) is no longer visible at the bottom of the tube.
- Place the tube on ice
- Label five tubes "Enzyme 1".
- Transfer 35 µl of diluted Restriction Enzyme 1 to each tube.
- Cap the tubes and immediately put on ice.

7. Dilute and Dispense Enzyme #2:
   - With a clean, dry 1 ml pipet, add 0.25 ml of **ice cold** Enzyme Dilution/Reaction Buffer (F) to Restriction Enzyme 2 (H).
   - Mix by gently tapping until the denser layer of glycerol solution (in the enzyme) is no longer visible at the bottom of the tube.
   - Place the tube on ice
   - Label five tubes "Enzyme 2".
   - Transfer 35 µl of diluted Restriction Enzyme 2 to each tube.
   - Cap the tubes and immediately put on ice.

You have now prepared the DNAs and restriction enzymes for each student group to perform four restriction enzyme reactions. Each student group should receive the following materials for the restriction enzyme digestion part of this experiment:

Reagents and biologicals listed above in Table 1
Automatic micropipet and yellow tips
4 microtest tubes with attached caps
Marking pen

One tube (1 ml) of 10x gel loading solution, included with this kit, should be shared by all five lab groups.

Method

Restriction Enzyme Digestion

1. Label microtest tubes 1 through 4 for four restriction enzyme digestion reactions. Put your initials or group number on the tubes.

2. Tap all the tubes on the lab bench to collect all the contents at the bottom of the tube.

3. Using an automatic micropipet, dispense 10 μl of Dilution/ Reaction Buffer (Rxn Buffer) to each of the four reaction tubes labeled 1 through 4.

4. Add DNA and enzyme to the reaction tubes as summarized in Chart 1 below. Use a FRESH micropipet tip for each transfer of DNA or enzyme.

Chart 1: Summary of Restriction Enzyme Digestion Reactions

	Reaction Tube	Reaction Buffer	DNA 1	DNA 2	Enzyme 1	Enzyme 2	Final Volume
Crime Scene Samples	Crime Scene DNA, ready for electrophoresis				X	--	45 μl *
	Crime Scene DNA, ready for electrophoresis				--	X	45 μl *
Suspect 1	1	10 μl	15 μl	--	15 μl	--	40 μl
	2	10 μl	15 μl	--	--	15 μl	40 μl
Suspect 2	3	10 μl	--	15 μl	15 μl	--	40 μl
	4	10 μl	--	15 μl	--	15 μl	40 μl

* 10x Gel loading solution has already been added to the crime scene samples.

5. Cap the reaction tubes and tap gently to mix. Then tap each tube on the lab bench to collect the contents at the bottom.

6. Incubate the four reaction tubes in a 37°C water bath for 30 minutes.

After the 30 minute incubation is completed

7. Add 5 μl of 10x gel loading solution to reaction tubes 1 - 4 to stop the reactions. Cap and mix by tapping.

Method, continued

Agarose Gel Electrophoresis

This experiment requires a 0.8% agarose gel with 8 sample wells. Have a water bath or beaker of water warmed to 65°C to heat samples before gel loading. Refer to Unit 1, Module 1 for information regarding gel preparation and running the gel.

Loading DNA Samples

1. Heat the samples, including the Standard DNA fragments for two minutes at 65°C. Allow the samples to cool for a few minutes.

2. Load DNA samples in the following manner:

First Row

Lane	Tube	
1	Markers	Standard DNA Fragments
2	CS 1	DNA from crime scene cut with Enzyme 1
3	CS 2	DNA from crime scene cut with Enzyme 2
4	1	DNA from Suspect 1 cut with Enzyme 1
5	2	DNA from Suspect 1 cut with Enzyme 2

Second Row

Lane	Tube	
1	Markers	Standard DNA Fragments
2	3	DNA from Suspect 2 cut with Enzyme 1
3	4	DNA from Suspect 2 cut with Enzyme 2

For further details about agarose gel electrophoresis, review the procedures outlined in Unit 1, Module 1.

Results

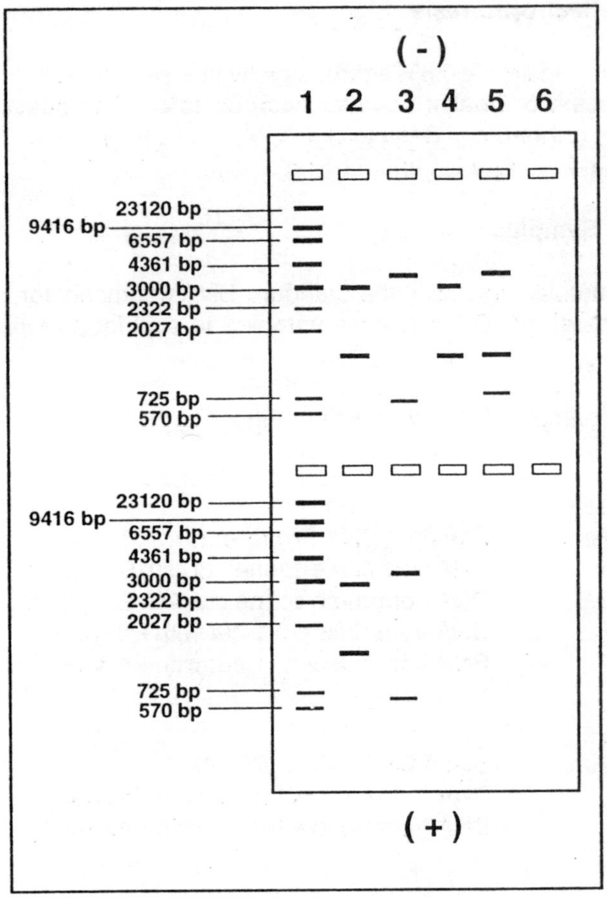

The figure to the left is an idealized schematic showing relative positions of DNA fragments. Actual results will yield broader bands of varying intensities. Smaller fragments will stain less efficiently and will appear as fainter bands. The idealized schematic shows the relative positions of the bands, but are not depicted to scale.

First Row

Lane	Tube	
1	Markers	Standard DNA Fragments
2	CS 1	Crime scene DNA cut with Enzyme 1
3	CS 2	Crime scene DNA cut with Enzyme 2
4	1	Suspect 1 DNA cut with Enzyme 1
5	2	Suspect 1 DNA cut with Enzyme 2

Second Row

Lane	Tube	
1	Markers	Standard DNA Fragments
2	3	Suspect 2 DNA cut with Enzyme 1
3	4	Suspect 2 DNA cut with Enzyme 2

UNIT 3

Bacterial Plasmid Transformation

UNIT 3: BACTERIAL PLASMID TRANSFORMATION

MODULE 10: ASEPTIC TECHNIQUES IN BACTERIAL CULTURING

Mark Petersen
Jordan Choper
Jack G. Chirikjian

* Introduction
* Safety Guidelines
* Experimental Outline
* Materials
* Pre-lab Preparation
* Method
* Results

Introduction

During the late 1800s, researchers worked to establish the cause of disease. Many suggested that disease was caused and spread by germs. This theory was supported by observations that rod-shaped microorganisms in the blood of animals caused anthrax. Pasteur showed that certain germs produced good wine, and other germs caused bad wine. Identification of the causative agent or germ was based on crude morphologic microscopic examination.

It was Koch who insisted on proof that an organism caused disease. Koch developed and refined techniques for the isolation of pure cultures by using solid agar medium. Koch was able to isolate the anthrax germ in pure culture by streaking onto his solid media. This isolated microorganism could be inoculated into animals and cause disease. Based on these experiments, Koch formulated criteria that provided proof that a specific microorganism caused disease. Known as Koch's postulates, the criteria are:

1. The organism must always be found in the diseased individual but not in healthy ones.
2. The organism isolated from a diseased individual must be grown in pure culture.
3. The organism isolate in pure culture must cause the disease when introduced into susceptible individuals.
4. The organism should be reisolated from the experimentally infected individual.

By 1900, most major bacterial disease organisms had been described. Today, clinical identification of a disease causing microorganism still relies heavily on streaking onto solid agar media for isolation of pure cultures. Isolation depends on the source of the specimen. Blood and spinal fluid do not normally harbor microorganisms, so isolation would be easier from blood or spinal fluid than from a throat swab. The throat hosts many microorganisms as normal flora. Disease-causing microorganisms (pathogens) may be present in small numbers compared to normal flora organisms. Specialized streaking techniques on solid agar media allow isolation of individual colonies. Individual colonies arise from a single organism (cell) which multiplies rapidly and forms macroscopic, visible masses of growth called colonies. Morphologic characteristics of colonies aid in identification of the pathogen.

Individual colonies of different organisms may vary in size, color, odor, border shape, and texture. Colonies can be 1 mm to 5 mm in size, with some organisms characteristically forming small colones (1 mm) such as *Hemophilus*. Some microorganisms produce pigments forming pink, yellow, or green colonies such as *Pseudomonas*. Organisms utilize different metabolic pathways resulting sometimes in gas production with distinctive and pungent odors. Borders of individual colonies can form unique shapes such as wavy, serrated or smooth. The interior of the colony may form

Introduction, continued

textures which appear mucoid (M) or water like, smooth (S) with a uniform texture, and rough (R) with a granulated appearance.

Upon isolation of single colonies, microscopic and biochemical characterization can be done. Biochemical tests can determine substrate utilization, metabolic product formation, sugar fermentation and antibiotic susceptibility. Microscopic examination can identify pathogens based on cell shape. Common shapes are cocci (spherical), bacilli (rodlike), and spiral. To better visualize microorganisms, special stains are used to detect outer membranes, capsules, flagella, spores, and intracellular inclusion bodies. Of all of the staining procedures for bacteria, most important is the Gram stain. Bacteria have been grouped based on their Gram stain reaction (positive or negative).

H. Christian Gram developed the Gram stain in 1883. The exact chemistry is still not fully understood. Cell wall permeability and the physicochemical differences between cell walls account for differential retention of stain in Gram-positive and Gram-negative organisms. Under electron microscopy, cross sections of gram-positive cells reveal a thick layer or peptidoglycan overlying the cytoplasmic membrane. The thick peptidoglycan layer can contain proteins and polysaccharide moietics which contribute to the layered structure. Gram-negative bacteria exhibit three distinct layers: the outer membrane, a middle dense layer and the inner cytoplasmic membrane. The middle dense layer contains a periplasmic space with a gel-like substance and only a few layers of peptidoglycan. Both the outer membrane and cytoplasmic membrane exhibit the typical lipid bilayer structure. Points of connection between the inner and outer membranes are known as adhesion sites.

The Gram stain involves a differential staining procedure utilizing several stains. The primary stain is crystal violet which is applied for 1 minute. Second, a solution of iodine is added for 1 minute as a mordant. Mordants react with the primary stain and cell components to enhance the retention of the primary stain. This causes purple insoluble complexes to form with the cell's RNA. The subsequent alcohol rinse causes Gram-negative bacteria to lose these complexes while Gram-positive organisms retain them. Loss or retention of these complexes is related to the permeability of the cell wall. The procedure is completed by counterstaining with safranin for 1 minute. This red stain is taken up by decolorized Gram-negative cells. Thus, Gram-negative cells are stained pink/red and Gram positive cells are stained purple/blue.

In this module, students will practice important microbiology methods, including isolation of a pure culture, utilizing sterile techniques. Once a single colony is isolated, colony size, odor, color and shape will be observed. After colony morphology has been observed, a Gram stain will be performed.

Safety Guidelines

1. Gloves and goggles should be worn routinely as good laboratory practice.

2. Exercise extreme caution when working with equipment which is used in conjunction with the heating and/or melting of reagents.

3. DO NOT MOUTH PIPET REAGENTS - USE PIPET PUMPS OR BULBS.

4. The two bacteria used in this experiment are not considered pathogenic, but all micoorganisms have the potential to cause disease in some individuals. Although the two bacteria are rarely associated with illness in healthy individuals, it is good laboratory practice to follow simple safety guidelines in handling and disposal. At the completion of the experiment:

   A. Wipe down the lab bench with a 10% bleach solution, disinfectant or soapy water.

   B. All materials, including petri plates, pipets, transfer pipets, loops and tubes, that come in contact with bacteria should be disinfected before disposal in the garbage. Disinfect materials as soon as possible after use in one of the following two ways:

      - Autoclave at 121°C for 20 minutes.
        Tape several petri plates together and close tube caps before disposal. Collect all contaminated materials in an autoclavable, disposable bag. Seal the bag and place it in a metal tray to prevent any possibility of liquid media or agar from spilling into the sterilizer chamber.

      - Soak in 10% bleach or disinfectant solution.
        Immerse petri plates, open tubes and other contaminated materials into a tub containing a 10% bleach solution. Soak the materials overnight and then discard. Wear gloves and goggles when working with bleach.

   C. Always wash hands thoroughly with soap and water after handling contaminated materials.

Experimental Outline

1. Preparation of agar plates
2. Preparation of stock organism plates
3. Gram staining of microorganisms
4. Preparation of bacterial slides
5. Examination of slides under microscope

Materials

This module utilizes EDVOTEK experiment #160, Principles and Practice of Microbiology (1-800-EDVOTEK). The following components are included in the kit:

Slant of *Micrococcus luteus*
Slant of *Serratia marcescens*
Gram's Crystal Violet
Gram's Iodine
Gram's Safranin

ReadyPour™ Luria Broth Agar
Sleeve petri plates (sterile)
10 ml pipet, sterile
Inoculating loops
Transfer pipets

Requirements

Incubation oven
Microscope with oil immersion lens
Microscope slides
Microscope oil
Slide holders or clothespins
95% ethanol
Bunsen burners
Lab markers
Toothpicks
Hot plate, bunsen burner or microwave
(for heating ReadyPour agar)
Gloves
Goggles
Lab coats
Bleach or disinfectant
Trays for disinfectant
25 ml flasks
Distilled water
Squeeze bottles

Pre-lab Preparation

General Information

1. Pre-lab preparations will require approximately one hour and 30 minutes. Stock organism plates should be prepared at least 2 days prior to the laboratory. Agar plates can be prepared up to a week in advance. Gram's stain reagents can be dispensed any time before the experiment.

2. Students require 30 minutes on Day One to streak plates and 40 minutes to 1 hour on Day Two to complete the Gram's stains.

Preparation of Agar Plates

Sterile agar plates can be prepared several days before the laboratory experiment. They should be stored in refrigerator prior to use.

1. Equilibrate a water bath at 50°C for step 5.

2. Loosen, but do not remove, the cap on the ReadyPour media bottle to allow for the venting of steam during heating.

3. Heat the bottle of ReadyPour media by one of the methods outlined below. The amber-colored solution should appear free of small particles.

   A. Water Bath or oven method:

   - Heat the bottle in a boiling (100°C) water bath or oven.
   - Using hot glove, occasionally swirl to expedite melting.

   B. Microwave method:

   - Heat the bottle on high for two 30 second intervals.
   - Swirl and heat on high for additional 25 seconds (until all ReadyPour media is dissolved).

   C. Hot plate or burner method:

   - Place the bottle in a beaker partially filled with water.
   - Heat the beaker to boiling over a hot plate or burner.
   - Using hot glove, occasionally swirl to expedite melting.

4. Cool the melted ReadyPour media to approximately 50°C.

5. Place the bottle in a 50°C water bath to prevent the agar from prematurely solidifying.

Pre-lab Preparation, continued

After the ReadyPour media has cooled, pour 16 plates as outlined in steps 6-10:

6. Use a sterile 10 ml pipet with a pipet pump to transfer 5 ml of media to a petri plate. Pipet carefully to avoid forming bubbles.
7. Rock the petri plate back and forth to obtain full coverage.
8. Cover the petri plate and allow the media to solidify.
9. If the solidified media contain bubbles, they can be removed by passing a flame across the surface of the solidified media.
10. Wrap the plates in foil, plastic wrap or parafilm and store them inverted, in the refrigerator.

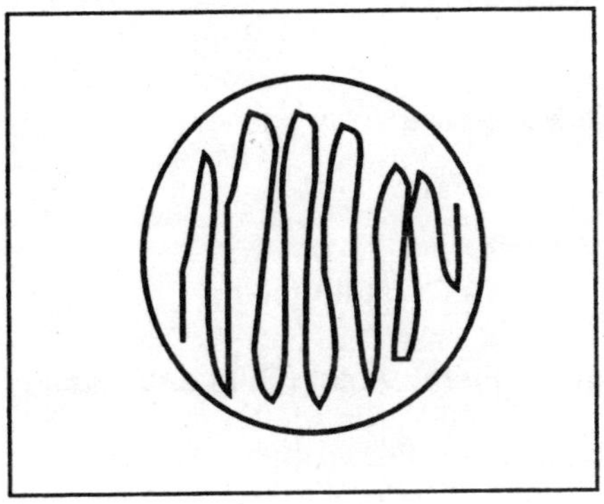

Figure 1

Preparation of Stock Organism Plates (2 Days before the lab)

1. Prepare a 10% solution of bleach or disinfectant by adding 10 ml bleach or disinfectant to 90 ml water. Prepare enough to provide 1/2 inch of solution to cover the bottom of a tray.
2. Equilibrate an incubation oven at 30°C for step 13 below.

Preparation of Micrococcus Luteus stock plates

3. Label 3 agar plates "Micrococcus Luteus".
4. Remove one sterile inoculating loop from the package.
5. Open the slant of *Micrococcus luteus* (A). Remove inoculum from the slant by passing the loop over the slant surface.
6. Spread the inoculum over the surface of one *Micrococus luteus* plate as shown in Figure 1. Dispose of the loop in the 10% bleach or disinfectant.
7. Repeat steps 4 - 6 for the other two *Micrococus luteus* plates.

Preparation of Serratia Marcescens stock plates

8. Label 3 agar plates "Serratia Marcescens".
9. Remove one sterile inoculating loop from the package.
10. Open the slant of *Serratia Marcescens* (B). Remove inoculum from the slant by passing the loop over the slant surface.

Pre-lab Preparation, continued

11. Spread the inoculum over the surface of one *Serratia marcescens* plate as shown in Figure 1. Dispose of the loop in the 10% bleach or disinfectant.

12. Repeat steps 10 - 11 for the other two *Serratia marcescens* plates.

Incubation of Stock Plates

13. Incubate all plates inverted at 30°C overnight.

14. After incubation, heavy growth should be evident. If growth is not heavy (very evident), incubate longer if possible.

15. Plates can be stored inverted in refrigerator until experiment.

Dispensing of Gram Stain Reagents

Gram's Crystal Violet

1. Label five small flasks "Gram's Crystal Violet".

2. Dispense 10 ml of Gram's Crystal Violet (C) to each flask.

3. Cover to prevent evaporation.

Gram's Iodine

4. Label five small flasks "Gram's Iodine".

5. Dispense 10 ml of Gram's Iodine (D) to each flask.

6. Cover to prevent evaporation.

Ethanol

7. Label five small flasks "95% ethanol".

8. Dispense 20 ml of 95% ethanol to each flask.

9. Cover to prevent evaporation.

Gram's Safranin

10. Label five small flasks "Gram's safranin".

11. Dispense 10 ml gram's safranin (E) to each flask.

12. Cover to prevent evaporation.

Pre-lab Preparation, continued

Assembly of Student Materials

For Day One, each lab group should receive:	
2	Agar plates
1	tray with 10% disinfectant or bleach
1	marker to label plates
2	sterile inoculating loops
	Gloves
	Goggles
	Labcoat

For Day Two, each lab group should receive:		
2	microscope slides	
1	Squeeze bottle filled with distilled water	
1	Bunsen burner	
2	toothpicks (or nichrome loop)	
4	Transfer pipets	
1	slide holder or clothespin	
1	Gram's Crystal Violet	10 ml
1	Gram's Iodine	10 ml
1	95% Ethanol	20 ml
1	Gram's Safranin	10 ml
	Microscope oil	
	Paper towels	
	Gloves	
	Goggles	
	Labcoat	

Method, continued

Day One Experimental Procedures

Prior to performing stain or biochemical tests, it is important to isolate single bacterial colonies. Heavy bacterial growth covering the entire surface of an agar plate could have more than one bacterial species. Removing a loop of bacterial growth containing several bacterial species, conducting stain or biochemical tests could yield inconclusive results. Because bacteria contaminate many surfaces, including human skin, touching the surface of an agar plate will produce bacterial growth. To avoid contaminating bacterial media, it is important to use sterile technique.

Streaking for Pure Cultures

1. Obtain 2 nutrient agar plates. Label one plate "Serratia marcescens" on bottom half. Label the second plate "Micrococcus luteus". Write your group number or initials on each plate.

2. Obtain a stock plate of *Serratia marcescens* that was prepared by your instructor.

Figure 1

Figure 2

3. Carefully remove one inoculating loop from the sterile pouch. Do not touch the loop end with your gloved hands or place the loop on the table surface. If the loop touches any surface, discard it in the disinfectant tray and obtain another sterile loop.

4. Open the lid of the *Serratia marcescens* stock plate just enough to insert the loop. Touch the loop to a colony or area of growth.

5. Remove the loop and replace the lid on the stock plate.

6. Open the lid of your nutrient agar plate labeled "Serratia marcescens" just enough to insert the loop, and streak as shown in Figure 1.

7. Rotate your plate a quarter turn. Streak once through the first streak area and continue streaking down the plate as shown in Figure 2. Do not touch the primary streak area as you streak down the plate.

8. Discard the loop in the disinfectant tray.

9. Repeat the same procedure (steps 3-8) with *Micrococcus luteus* on the second nutrient agar plate.

10. Incubate both plates inverted (agar side up) at 30°C, or at room temperature for 24 to 48 hours as indicated by your instructor.

11. Growth should be evident on both plates. If growth is minimal, incubate longer.

Method, continued

Day Two Experimental Procedures

Observation of Plates

1. Look for growth on the surface of each plate. Lift lid only enough to make observations. Do not remove lid. The primary streak area should have heavy growth. The second streak area should have heavy growth at the top and individual colonies.

2. Record your observations concerning the following parameters for each plate:

   - colony size
   - color or pigment
   - border shape (wavy, serrated or smooth)
   - texture (smooth, rough or mucoid)

3 To check for odor, lift the lid half an inch. Slowly fan your hand toward your nose. Do not put the plate near your face. Try to describe the odor, if any.

4. Now perform your Gram stain of each organism.

Gram's stain

1. Light a bunsen burner.

2. Write #1 and your initials or group number on a microscope slide.

3. Using a slide holder or clothespin, pass the microscope slide through the flame twice.

4. Lay it on a paper towel to cool.

5. Write #2 and your initials or group number on a second slide.

6. Pass it through the flame twice with the slide holder.

7. Lay it on a paper towel to cool.

8. Place one drop of distilled water on slide #1.

9. Raise the lid of your nutrient agar plate labeled "Serratia marcescens". Touch a single colony with a toothpick.

10. Stir the toothpick through the drop of distilled water. The liquid should look turbid.

Method, continued

11. Discard the toothpick in the disinfectant tray.

12. Allow the water drop to air dry until completely dry.

13. Repeat steps 8-12 for slide #2 using *Micrococous luteus.*

After the slides have completely dried:

14. Pass the dried smear of each slide through the flame twice, smear side up. Do not hold the slides in the flame.

15. Turn off the bunsen burners. Allow the slides to cool.

16. Using a transfer pipet, place 3 to 4 drops of Gram's crystal violet to cover the smears on each slide. Time for 20 seconds.

17. Rinse with slow running tap or distilled water.

18. Place 3 to 4 drops of Gram's iodine on smears. Time for 20 seconds.

19. Rinse with slow running tap or distilled water.

20. Hold the slide at an angle. Using a transfer pipet, gently flush the smears with 95% ethanol for 5 seconds. Rinse immediately with tap or distilled water.

21. Shake off excess water and add 3 to 4 drops of Gram's safranin to each smear. Time for 40 seconds.

22. Rinse the slide using a squeeze bottle filled with distilled water or under a running tap.

23. Gently blot water from the slide with a paper towel. Do not rub.

24. These slides can be stored in a drawer at room temperature. Do not stack them on top of each other.

Examining the slides under a microscope

1. On the lowest power, focus on the color smear.

2. Switch to next magnification. Again focus on colored smear.

3. Continue until the next lens is the oil immersion lens.

4. Place a drop of microscope oil on the smear.

5. Lock the oil immersion lens in place.

Method, continued

6. Slowly raise the stage until the lens enters the oil.

7. Adjust the focus until the bacterial cells are in focus.

8. Examine your gram stain for:

   - Shape (cocci or bacilli)
   - Dye retention (color)

Optional Activity

1. Light a bunsen burner.

2. Write "Mix" and your initials or group number on a microscope slide.

3. Using a slide holder or clothespin, pass the microscope slide through the flame twice.

4. Lay it on a paper towel to cool.

5. Place two drops of distilled water on the slide approximately 1/2 inch apart.

6. Raise the lid of your nutrient agar plate labeled "Serratia marascens". Touch a single colony with a toothpick.

7. Stir the toothpick through a drop of distilled water. The liquid should look turbid.

8. Discard the toothpick in the disinfectant tray.

9. Raise the lid of your nutrient agar plate labeled "Micrococcus luteus". Touch a single colony with a toothpick.

10. Stir the toothpick through the other drop of distilled water.

11. Using the same toothpick, mix the two drops together.

12. Discard the toothpick in the disinfectant tray.

13. Allow the water drop to air dry until completely dry.

14. Using a slide holder, pass the dried smear through the flame twice, smear side up. Do not hold the slide in the flame.

15. Turn off the bunsen burners. Allow the slide to cool.

16. Repeat the steps for Gram stain.

Results

Serratia marcescens

- colony size: 3-5 mm
- color: pink to red
- border: smooth
- texture: smooth to mucoid
- strong odor present
- slide #1: Gram-negative bacilli (pink to red)

Micrococcus luteus

- colony size: 1-2 mm
- color: white to yellow
- border: smooth
- texture: smooth
- light odor present
- slide #2: Gram-positive cocci (purple to blue)

Typical Results for Optional Activity

Slides show Gram-negative bacilli and Gram-positive cocci. Color differences should be evident between Gram-negative and Gram-positive.

UNIT 3: BACTERIAL PLASMID TRANSFORMATION

MODULE 11: TRANSFORMATION OF *E. coli*

Ellie Skokan

* Introduction
* Safety Guidelines
* Experimental Outline
* Materials
* Pre-lab Preparation
* Method
* Results

Introduction

The objectives of this module are to:

1. Demonstrate the change in genetic makeup (transformation) of a bacterial cell, using a plasmid vector.
2. Demonstrate the use of a phenotypic marker to determine the success of transformation.
3. Calculate the efficiency of the transformation process.

Transformation may be described as the stable, heritable uptake of exogenous (i.e., from another source, or so-called, foreign) DNA into a host cell. The source of the DNA in this experiment is a plasmid and the host cell is a bacterium.

Some genera of bacteria undergo transformation naturally. Natural transformation serves to add genetic variability to the population. Cells with the ability to undergo natural transformation are said to be naturally competent. In order for a host cell to undergo transformation, it must become competent.

Other bacteria, such as *E. coli*, may be artificially induced to become competent and then can be transformed, i.e., they can be made artificially competent. The exact mechanism of DNA uptake by cells with artificially induced competency is unknown, but it is known that the presence of cations, such as calcium or magnesium, is required. It is also known that transformation can be enhanced by cycles of hot and cold temperatures.

In transformation attempts, competent host cells are exposed to exogenous DNA. Then the host cells are subjected to optimum physical conditions for growth to allow the cells to recover (known as recovery). The success of transformation is measured in terms of transformation efficiency. Transformation efficiency is the number of host cells transformed per microgram of DNA and is calculated as follows:

$$\frac{\text{number of transformed cells}}{\mu\text{g of DNA used}} \times \frac{\text{final volume of cell suspension (ml)}}{\text{volume of cell suspension used (ml)}} = \text{number of transformants per } \mu\text{g DNA}$$

For example, assume that by using 20 ng (0.02 μg) of DNA in a final volume of 2 ml, 50 cells were found to be transformed when a 0.1 ml aliquot of the final volume was tested. The transformation efficiency would equal 5×10^4 transformations per μg DNA.

Introduction, continued

$$\frac{50 \text{ cells transformed}}{0.02\ \mu\text{g DNA}} \times \frac{2 \text{ ml}}{0.1 \text{ ml}} = 50{,}000\ (5 \times 10^4)$$

The objective of the experiment is to attempt to transform competent cells of *E. coli* with the plasmid pGAL™. The *E. coli* used in this experiment does not produce restriction enzymes, lacks plasmids, is sensitive to the antibiotic ampicillin and lacks a functional gene for the production of the enzyme beta-galactosidase which is capable of hydrolyzing an analog of lactose, known as X-GAL (5-bromo-4-chloro-3-indolyl-beta-D-galactopyranoside). The pGAL™ plasmid, on the other hand, contains the gene for production of beta-lactamase (an enzyme which destroys ampicillin) and the gene for beta-galactosidase production. X-GAL is colorless, but the product formed when it is hydrolyzed by beta-galactosidase is blue in color. If transformation is successful, i.e., if the host cells take up the plasmid, the transformed cells will be resistant to ampicillin and will convert colorless X-GAL into its blue product.

In this experiment, you will mix a sample of plasmid with competent cells. The mixture will be subjected to optimum conditions for transformation followed by optimum conditions for recovery. In order to determine the success of transformation, the cells will be transferred to agar media containing a combination of X-GAL and ampicillin. The host cells which have been transformed should appear as blue colonies, indicating resistance to ampicillin and the presence of X-GAL hydrolysis.

The control for the experiment consists of treating competent cells with a buffer (no plasmids present) and subjecting the mixture to the same conditions as the test mixture. Cells from the control will be plated on agar containing X-GAL (to show the lack of the beta-galactosidase gene) and on agar containing ampicillin and X-GAL (to show the lack of the beta-lactamase gene).

Safety Guidelines

Use protective gloves when handling hot agar and/or dry ice.

At the completion of the experiment, materials exposed to bacteria should be collected in metal trays or other containers suitable for autoclaving during the laboratory periods. At the completion of this experiment, all materials should be sterilized for 20 minutes before disposal. Alternatively, the plates and all materials exposed to the cells, can be soaked in 10% bleach overnight and then discarded.

Experimental Outline

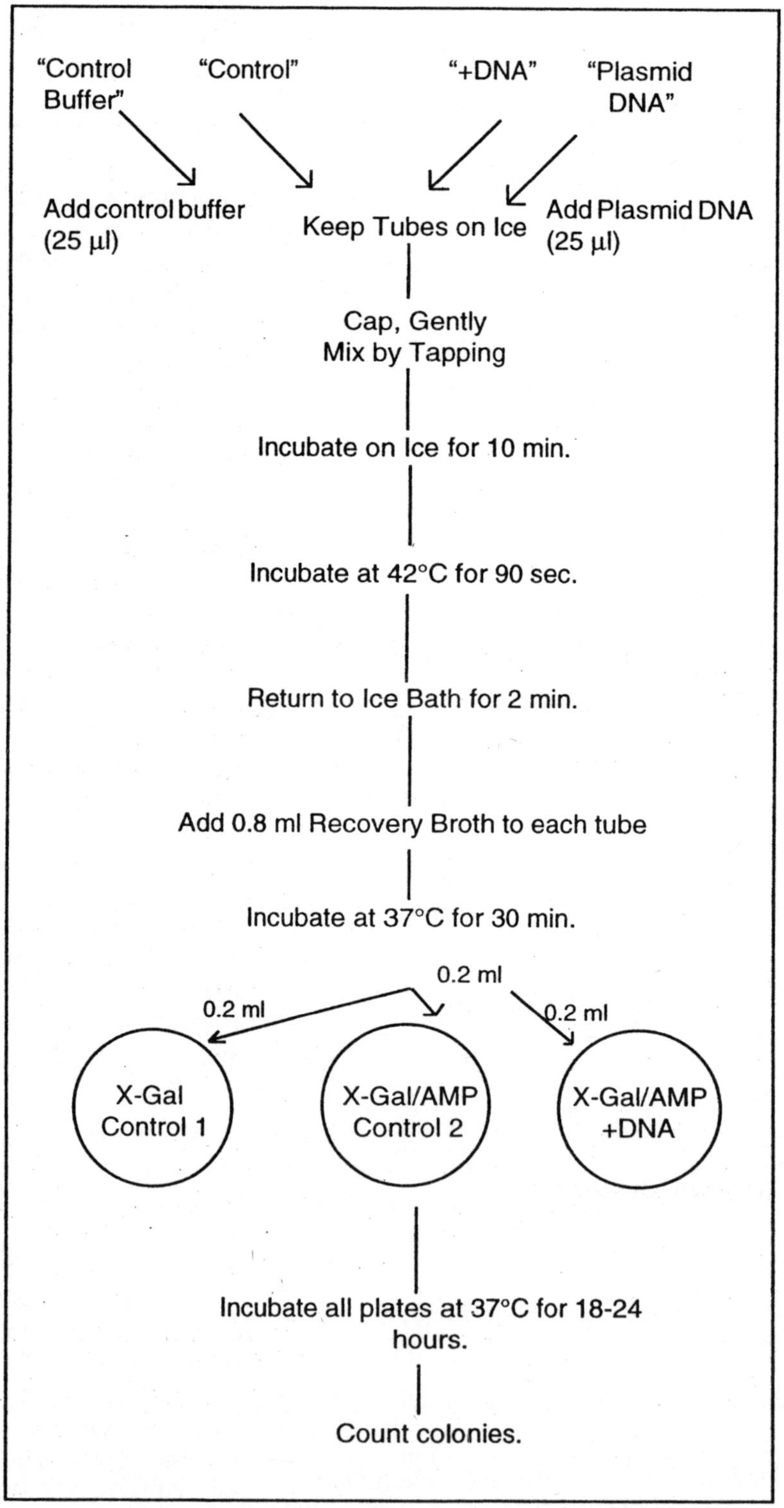

Figure 1: Transformation of *E. coli* with Plasmid pGal™

Materials

Each student group will need the following:

1 tube (0.3 ml) competent cells labeled "+DNA"
1 tube (0.3 ml) competent cells labeled "Control"
1 microtest tube (25 μl) labeled "Plasmid DNA"
1 microtest tube (25 μl) labeled "Control Buffer"
1 petri dish agar labeled "X-Gal"
2 petri dishes agar labeled "X-Gal/AMP"
Luria broth
2 disposable micro transfer pipets
3 sterile l ml pipets and pipet pump or bulb
2 disposable inoculating loops (or one regular and Bunsen burner)
Ice bath
37°C water bath
42°C water bath
37°C incubator
Gloves and goggles
Marking pen/pencil, waterproof ink
Container of disinfectant for disposal of used tubes/vials/pipets/
Loops

Items supplied as part of EDVOTEK Kit #221 (1-800-EDVOTEK):

Presensitized, LyphoCells™ (DO NOT FREEZE)
Supercoiled pGAL™ (Blue Colony) DNA
Control Buffer (no DNA)
Ampicillin (23 mg)
X-Gal in solvent (pre-measured)
Cell reconstitution media
Solvent for induction of competency
ReadyPour Luria Broth Agar, sterile
(also referred to as ReadyPour media)
Luria Broth Media for Recovery, sterile
(also referred to as Recovery Broth)
Petri plates (sterile)
Plastic microtipped transfer pipets
Wrapped 10 ml pipet (sterile)
Wrapped 1 ml pipets (sterile)
Inoculating loops (sterile)
Microtest tubes with attached lids

Materials, continued

Requirements

Micropipet (automatic or capillary capable of delivering 25 µl)
Two water baths (37°C and 42°C) NOTE: 42°C bath needed for only a few minutes (water may be heated to temperature in a beaker if used immediately)
Racks to hold tubes in water baths
Incubator/Oven, 37°C
Pipet pumps or bulbs
Ice
Marking pens, water proof ink
Protective gloves
Goggles
Disposable gloves
Containers for collection of used materials for disposal

Pre-lab Preparation

Timetable of Events

DNA, buffer solution, ampicillin and X-Gal.

1-3 days before lab: Prepare agar plates (approx. 4 hours). Store at room temperature inverted.

1-4 hours before lab: Thaw DNA and buffer on ice.

1 hour before lab: Thaw competent cells. Aliquot DNA, buffer and cells.

Students can complete experiment in 2 hour period. Second 2 hour period necessary to collect data, discuss results and answer study questions (advanced students will complete in less time.)

Preparation of Agar Plates

1. If using EDVOTEK Experiment #221, follow directions included and proceed to Reagent Preparation.

2. If preparing your own media, prepare sufficient Luria Broth Agar (15 g/L agar, 20 g/L Luria Broth Growth Medium) for 30 plates. After sterilization, cool the agar to 50°C. Maintain the agar at 50°C until the plates are prepared.

3. Add all of the X-Gal solution to the agar. Recap the container and swirl to mix.

Pre-lab Preparation, continued

4. Pour 5 ml agar containing X-Gal into 10 sterile petri dishes.

5. Add the ampicillin to the remaining agar. Recap the container and swirl to mix. Pour the agar containing X-Gal and ampicillin into 20 sterile petri dishes.

6. Briefly pass a flame over the surface of the media to remove bubbles. Cover the plates and let the media solidify.

7. Label the plates containing agar and X-Gal "X-Gal". Label the plates containing agar, X-Gal and ampicillin "X-Gal/AMP".

8. Wrap the plates in foil, plastic wrap or parafilm and store them, inverted, in the refrigerator until used. The plates should be used within 2-3 days.

Preparation of Competent Cells

1. Day before Lab

   - Use a 10 ml sterile pipet to add 6 ml sterile Cell Reconstitution media to the vial of LyphoCells. Save pipet for step 8.
   - Replace rubber stopper and cap. Mix by inverting until the freeze dried plug is dissolved.
   - Shake cell suspension vigorously and incubate vial at 34°C for 16 - 24 hours (overnight) in an incubation oven

2. Day of Lab:

   - Place competency induction solvent on ice.
   - Mix to resuspend incubated cells by inverting and gently shaking.
   - Place the vial on ice for 10 minutes.
   - Mix to resuspend cells.
   - With the pipet saved from step 1, add 3 ml of ice cold competency induction solvent.
   - Mix thoroughly by inverting several times. The solution should have no dense layers, "streams" or globules.
   - Keep the cells on ice for a minimum of 30 minutes.

Pre-lab Preparation, continued

3. Dispensing the cells just prior to the experiment (keep cells on ice during dispensing procedures):

   - Mix the cells by inversion to obtain an even suspension.
   - Use a sterile 1 ml pipet to aliquot 0.3 ml of cells to:

     10 ice cold tubes labeled "Cells for DNA"
     10 ice cold tubes labeled "Cells for Control" .

   - Cap the tubes and keep them on ice.

Preparation of Biologicals and Reagents

1. Make appropriate arrangements for incubation of LyphoCells prior to experiment.
2. On day of lab, thaw supercoiled pGAL™ DNA and buffer solution and place on ice.
3. Allow ample time for equilibration of water baths and incubation oven.

Preparation of DNA

1. Label ten microtest tubes “Plasmid DNA”.
2. Tap the tube of supercoiled pGAL™ DNA until all the sample is in the tapered bottom of the tube.
3. Using a micropipet, dispense 25 µl (0.025 ml) of the supercoiled pGAL™ DNA into each of the microtest tubes labeled “Plasmid DNA”.
4. Cap the tubes and place them back on ice.

Preparation of Control Buffer

1. Label ten microtest tubes “Control Buffer”.
2. Tap the tube of control buffer until all the sample is in the tapered bottom of the tube.
3. Using a micropipet dispense 25 µl (0.025 ml) of the control buffer into each of the microtest tubes labeled “Control Buffer”.
4. Cap the tubes and place them back on ice.

Method

1. Remove two tubes each containing 0.3 ml of competent cells (one labeled "(+)DNA" and one labeled "Control") from ice bath. The tube labeled "(+)DNA" will be treated with plasmids while the tube labeled "Control" will be treated with a buffer containing no plasmids. Obtain one microtest tube labeled "Plasmid DNA" and one microtest tube labeled "Control Buffer". Place all the tubes and microtest tubes in a beaker of ice.

2. Open the tube of cells labeled "(+)DNA" and the microtest tube labeled "Plasmid DNA". Using a fresh transfer pipet, slowly draw all of the solution in the tube labeled "Plasmid DNA" into the pipet and add all of the solution directly to cell suspension in tube labeled "(+)DNA". Briefly submerge and swirl the pipet tip in cell suspension when adding DNA solution.

   Cap the tube and gently mix by tapping two times. The amount of DNA added is 25 ng (0.025 μg). Put the tubes of cells back on ice.

3. Open the tube of cells labeled "Control" and the microtest tube labeled "Control Buffer". Using a fresh transfer pipet, slowly draw all the solution in the tube labeled "Control Buffer" into the pipet and add all the solution directly to the cell suspension tube labeled "Control". Briefly submerge and swirl pipet tip in cell suspension when adding the buffer solution. Cap the tube and gently mix by tapping two times. Put the tube of cells back on ice.

4. Let both tubes containing the cells you have prepared incubate on ice for 10 minutes.

5. Remove tubes from ice and incubate at 42°C for 90 seconds.

6. Remove the tubes from the 42°C water bath and put the tubes back on ice for two minutes.

7. Using a sterile l ml pipet, add 0.8 ml of Luria broth growth medium (known as recovery broth) to the tube labeled "Control". Do not touch the cell suspension with the pipet. Using the same pipet, add 0.8 ml of recovery broth to the tube labeled "+DNA". Do not touch cell suspension with the pipet.

   Cap both tubes and place in a 37°C water bath for 30 min.

8. Remove both tubes from the water bath.

Method, continued

9. Obtain three petri dishes containing agar, one labeled "X-Gal" and two labeled "X-Gal/AMP" and label as follows:

   a. Label the bottom of plate labeled "X-Gal" with "Control 1, your name/number".

   b. Label the bottom of one plate labeled "X-Gal/AMP" with "Control 2, your name/number".

   c. Label the bottom of the other plate labeled "X-Gal/AMP" with "+DNA, your name/number".

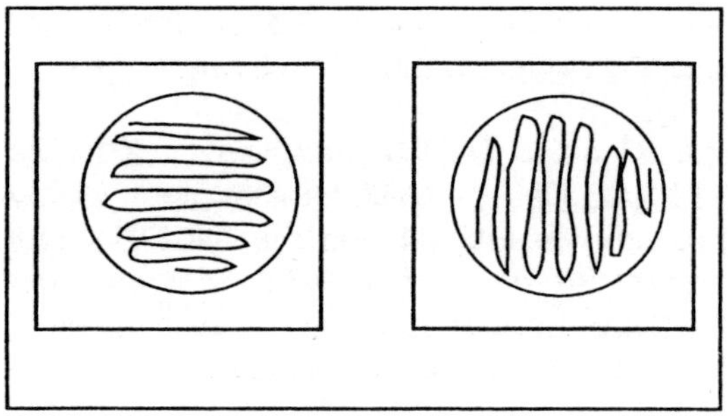

Figure 2

10. Using a sterile l ml pipet, transfer 0.2 ml of cells from the tube labeled "Control" to the center of the petri dish plate labeled "X-Gal, Control 1" and 0.2 ml of cells from the same tube to the center of the petri dish labeled "X-Gal/AMP, Control 2".

    Using a sterile inoculating loop, spread cells evenly over the entire surface of the agar in control plate as shown in Figure 2. Replace the lids on the culture plates.

11. Using a fresh, sterile l ml pipet, transfer 0.2 ml of cells from the tube labeled "+DNA" to the center of the petri dish labeled "X-Gal/AMP, +DNA". Using a sterile inoculating loop, spread the cells evenly over the entire surface of the agar as shown in Figure 2. Replace the lid on the culture plate.

12. Allow the agar to absorb some of the liquid (15-30 min). Invert all three agar plates and incubate at 37°C for 18-24 hours.

13. Estimate the number of colonies on the plate labeled "X-Gal/AMP, +DNA". If there are both blue and white colonies, estimate the number of each. Since each blue colony theoretically represents a single transformed cell, the number of blue colonies is equal to the number of transformed cells in 0.3 ml of cell suspension.

14. Calculate the transformation efficiency. Recall that transformation efficiency is the number of transformed cells per microgram of DNA. The quantity of DNA used in this experiment was 25 ng (0.025 μg). The final volume of the cell suspension in the tube was 1 ml and the volume of cell suspension used in plating was 0.2 ml.

15. Record the results for the plates labeled "Control 1" and "Control 2" and explain the results.

Results

Number of blue colonies/plate
X-Gal/AMP, +DNA:____________________
(= number of transformed cells/0.2 ml of cell suspension)

Number of white colonies/plate
X-Gal/AMP, +DNA:____________________

Calculated Transformation Efficiency:______________
transformants/μg DNA

Number of blue colonies/plate Control 1
(X-Gal):____________________

Number of white colonies/plate on Control 1:____________

Number of blue colonies/plate Control 2
(X-Gal/AMP):____________________

Number of white colonies/plate Control 2:____________

X-Gal Control 1	X-Gal/AMP Control 2	X-Gal/AMP +DNA
	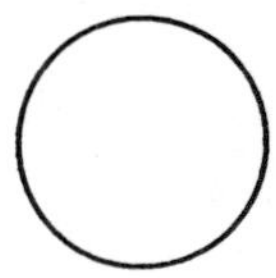	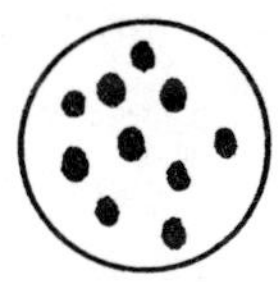
White colonies	No growth	Blue colonies
Plated with Control cells (no DNA)	Plated with Control cells (no DNA)	Plated with cells plus DNA
Plate will be covered with white colonies.	No growth	Blue colonies. White satellite colonies may be present.
Colonies are white because cells lack a functional gene for the production of beta-galactosidase and do not utilize X-Gal. Since no ampicillin is present, all cells should grow.	Cells lack ability to produce beta-lactamase and therefore the cells are ampicillin sensitive.	Demonstrates uptake of plasmid by host cells. Blue color indicates hydrolysis of X-Gal. Growth indicates resistance to ampicillin. Satellites are non-transformed cells (white) growing in area where ampicillin is inactivated.

UNIT 3: BACTERIAL PLASMID TRANSFORMATION

MODULE 12: PREPARATION OF COMPETENT CELLS FOR TRANSFORMATION

Harley Mortensen

* Introduction
* Experimental Outline
* Safety Guidelines
* Materials
* Method
* Results

Introduction

The objectives of this module are to produce *E. coli* competent cells suitable for transformation, transform the competent cells with a plasmid vector, and to screen the transformants for antibiotic resistance. Review Module 11 of this unit, Transformation of *E. coli,* for a discussion of transformation. This protocol differs from Module 11 in that the production of competent cells is included, and the transformation buffer is more complex.

Certain strains of *E. coli* can be rendered competent by appropriate treatment with certain cations such as Ca^{++}, Mn^{++}, Co^{+++}, K^{+}. 100mM $CaCl_2$ is commonly used for this purpose. The exercise described in this module is an alternative in which a transformation buffer containing potassium chloride, manganese II chloride, hexamine cobalt III chloride is used.

Experimental Outline

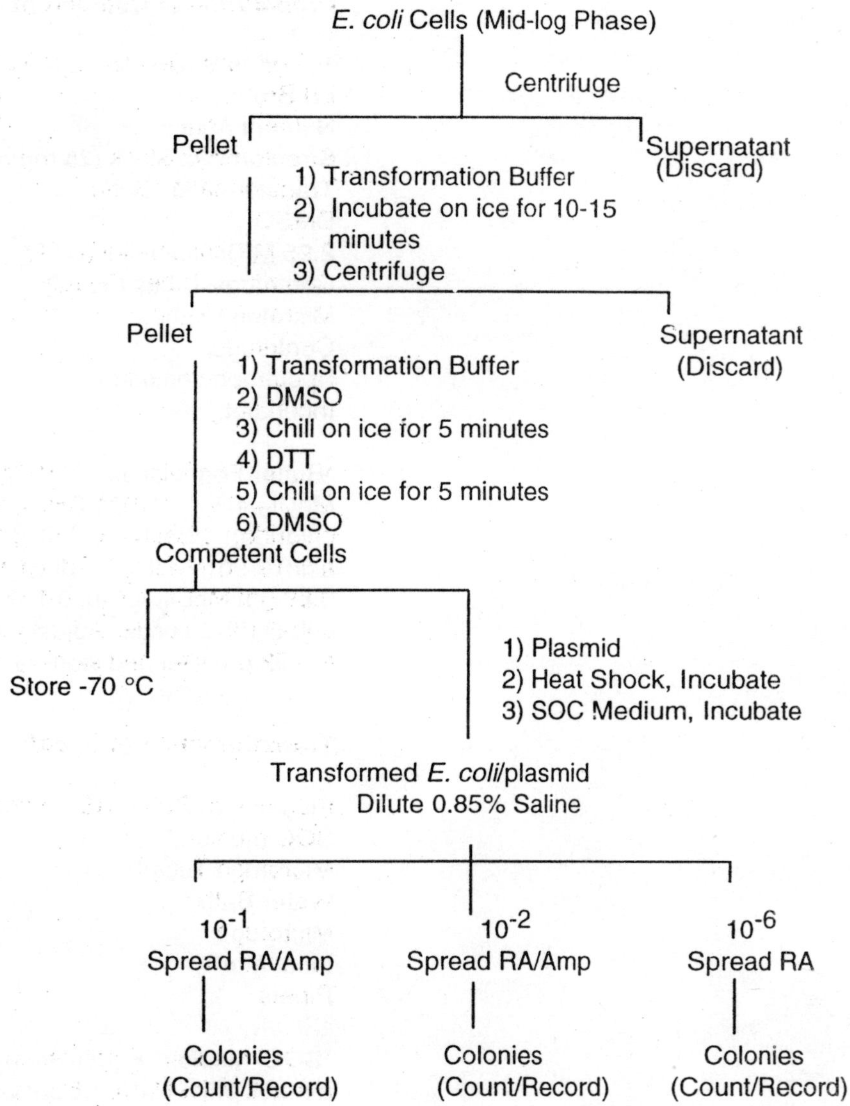

Safety Guidelines

Dimethyl Sulfoxide (DMSO) is readily absorbed through the skin. Use gloves when handling DMSO and ethidium bromide. Ethidium bromide is a powerful mutagen. Use goggles or other proper eye protection when viewing gels through a transilluminator. Ultraviolet radiation from the transilluminator can cause tissue damage.

Materials

Preparation of Competent Cells

E. coli cells (DH115) Life Technologies, Inc.
LB Broth
Nutrient Agar
Streptomycin Stock (25 mg/ml)
Transformation Buffer*
DMSO
2.25 M Dithiothreitol (DTT)
Centrifuge Tubes (50 ml)
Microfuge Tubes
Centrifuge
Spectrophotometer
Incubator

*Buffer Formulation: (0.01M K-MES (pH 6.2), 0.1M KCl, 0.045M $MnCl_2 \cdot 4H_2O$, 0.01M $CaCl_2 \cdot 2H_2O$, 0.003M hexamine cobalt (III) chloride): Dissolve 0.268 gm MES (2-morpholinoethanesulfonic acid) in 90 ml ddH_2O. Adjust pH to 6.2 with KOH. Add 0.745 gm KCl, 0.89 gm $MnCl_2 \cdot 4H_2O$, 0.147 gm $CaCl_2 \cdot 2H_2O$, 0.08 gm hexamine cobalt (III) chloride. Adjust volume to 100 ml, filter sterilize through a 0.22 μm filter and store at 4°C.

Transformation of *E. coli*

Plasmid pTZ18U (US Biochemical)
SOC medium*
Microfuge Tubes
Water Bath
Microfuge
Incubator
Pipets

*SOC Medium Formulation: 2% Bacto-tryptone, 0.5% yeast extract, 0.5% NaCl, 1.5% KCl. Autoclave. Bring to 20mM in Mg++ and 20 mM in glucose. (Filter sterilize a 2 M stock of 1 M $MgCl_2 \cdot 6H_2O$ + 1 M $MgSO_4 \cdot 7H_2O$. Filter sterilize a 2M glucose solution.)

Screening of Cells Transformed

0.85% saline (sterile)
Ampicillin Stock (50 mg/ml)
1 LB Plate
2 LB/Ampicillin (50 μg/ml) Plates
Incubator
Bunsen Burner
Glass Spreaders

Methcd

Preparation of Competent Cells

1. Prepare an overnight culture of cells by inoculation with a single colony taken from a LB agar/streptomycin (25 μg/ml) plate.

2. Transfer 100 μl of the culture to 50 ml of LB broth to which streptomycin has been added at 25 μg/ml.

3. Grow to early or mid-log phase (O.D. 350). The cells are to be chilled throughout the procedure and must be handled gently due to the fragility caused by the treatment required to make them competent.

4. Centrifuge at 7500 rpm for 10 minutes at 4°C.

5. Pour off the supernatant and add 15 ml of cold transformation buffer to the pellet. Agitate gently to suspend the cells and incubate on ice for 10-15 minutes.

6. Centrifuge at 7500 rpm at 4°C, decant the supernatant, and resuspend the cell pellet in 1 ml of transformation buffer.

7. Add 35 μl of dimethylsulfoxide (DMSO) and chill on ice for 5 minutes.

8. Add 35 μl 2.25 M dithiothreitol (DTT) and chill on ice for 10 minutes.

9. Repeat step 7. At this point, aliquots of 200 μl can be taken for immediate use or quick frozen and stored at -70°C. Competent cells stored in this manner can be used for several months with slight loss of competency with time.

Method, continued

Transformation of *E. coli*

1. Transfer 200 µl of the competent cells prepared above (or thaw one of the 200 µl aliquots) to a 1.5 ml Eppendorf tube.

2. Add 10 µg of plasmid or other DNA by placing the pipet tip at the bottom of the tube containing the cells and, while slowly drawing the pipettor out of the tube, expel the DNA evenly into the cell suspension. Do not shake or mix the suspension. Incubate the mixture on ice for 30 minutes.

3. Heat shock the mixture by suspending in a water bath at 42°C. for 1-2 minutes.

4. Incubate on ice for 10 minutes and then add 800 µl of SOC medium.

5. Incubate at 37°C. for one hour to allow time for gene expression.

Screening of Cells Transformed

Many useful vectors contain the gene for beta-lactamase which confers resistance to beta-lactam antibiotics such as ampicillin. The plasmid, pTZ18U, is one of these. Thus, this host/plasmid system can be routinely maintained by forcing retention of the plasmid by growing cells in the presence of ampicillin.

Cells will have been transformed according to procedures described in this module or in Module 11. The simple screening principle is that competent cells which are amp^s are rendered amp^r through uptake of plasmid DNA which carries the gene for ampicillin resistance.

1. After the one-hour incubation period called for in the transformation procedure, using sterile 0.85% saline as diluent, prepare 1 ml each of 10^{-1}, 10^{-2}, and 10^{-6} dilutions of the cell suspension.

2. Spread 100 µl of the 10^{-6} dilution on the LB plate. This is a control plate to determine the number of viable cells in the suspension.

3. Also spread 100 µl of 10^{-1} and 10^{-2} dilutions on LB/ampicillin plates. These allow a determination of the number of transformants in the suspension. Two dilutions are used in order to increase the likelihood of obtaining a countable plate.

4. Incubate all plates at 37°C.

Results

1. If colonies do not show up using the dilutions called for in item 3, spread 200 µl of undiluted cells on another LB/ampicillin plate and incubate at 37°C. This modification will frequently allow detection of transformants when the transformation efficiency is very low.

2. When colony growth has occurred sufficiently, count the colonies on each plate and calculate the cell concentration of each type that was present in the original culture. For example, if the control plate contained 15 colonies the calculation would be:

$$\frac{15 \text{ colonies (cells)} \times 10^6 \text{ (dilution factor)}}{0.1 \text{ ml}} = 1.5 \times 10^8 \text{ cells/ml}$$

3. Typical transformation efficiencies for students performing the exercise for the first time will be of the order of 10^4-10^5 cells per µg of DNA. Calculate the transformation efficiency for this exercise. For example, if 10 nanograms (.01 µg) of DNA were used in a 1 ml cell volume and if 100 µl of 10^{-1} cell dilution were plated and, after allowing time for colonies to appear, fifty colonies were present:

$$\frac{50 \text{ colonies (transformants)} \times 10 \text{ (dil factor)}}{0.01 \text{ µg DNA-1 ml suspension}} = \frac{5 \times 10^4 \text{ transformants}}{\text{µg DNA}}$$

4. Another way to examine the results: Calculate the ratio of the concentration of cells that are transformed to the concentration of total viable cells.

UNIT 4

Mutagenesis

UNIT 4: MUTAGENESIS

MODULE 13: SPONTANEOUS MUTATION - DIRECT SELECTION

Audrey M. Brown

* Introduction
* Safety Guidelines
* Materials
* Pre-lab Preparation
* Method
* Results

Introduction

The purpose of this unit is to differentiate between spontaneous and induced mutations. In addition, students learn the principles of direct (positive) and indirect (negative) selection techniques.

In this module, a spontaneous, streptomycin-resistant mutant of *E. coli* is isolated by direct selection using the gradient plate method. In Module 14, a spontaneous, streptomycin-resistant mutant of *E. coli* is isolated via indirect selection and replica plating.

A mutation is any change in the nucleotide sequence of DNA. These changes are heritable. Although most mutations are harmful to the organism, a small percentage are beneficial under certain environmental conditions. Mutations can occur during chromosome duplication and they may be spontaneous or they may be induced by mutagenic agents. Spontaneous mutations are thought to occur without any known cause. The rate of spontaneous mutation varies with each gene. In other words, some genes mutate at a faster rate than others. The rate of spontaneous mutation ranges from one in every one thousand cell divisions (1/1000) to one in every one billion cell divisions (1/1,000,000,000). The average rate is one in every one million cell divisions (1/1,000,000). A mutagenic agent or mutagen is any substance that increases the rate above that occurring spontaneously. Examples of mutagenic agents include certain chemicals, mutagenic viruses, and radiation. In this experiment, we will select for a spontaneous mutation causing streptomycin resistance in *E. coli*. Resistant mutants will be detected directly using the gradient plate method.

A gradient plate consists of two wedge-like layers of media. The bottom wedge is composed of plain nutrient agar. The top wedge consists of nutrient agar to which an antibiotic has been added. In this case, the antibiotic is streptomycin. The antibiotic located in the top wedge-like layer tends to diffuse into the plain agar. This creates a gradient of antibiotic concentration with one end containing a very low concentration and the opposite end possessing a high concentration.

We will select for spontaneous mutants of *E. coli* that are streptomycin resistant. Selection is accomplished by inoculating a gradient plate containing a high concentration of 100 μg/ml of streptomycin with a sensitive strain of *E. coli*. Colonies growing in the high concentration area are streptomycin resistant mutants.

Safety Guidelines

Students should be trained in basic microbiology techniques including but not restricted to: aseptic techniques, spread plating, pipeting procedures and discarding, and proper handling and transfer of bacterial cultures.

Used plates, bacterial cultures, pipets and contaminated materials should be properly autoclaved before discarding in a waste container.

Materials

Materials have been calculated for one pair or group of students.

Materials required for gradient plate preparation:

1 empty, sterile petri plate
2 nutrient agar deeps (10 ml per tube)
1% streptomycin solution (in water)
1 sterile 1 ml pipet
1 canister containing 95% isopropyl alcohol (or ethanol) for pipet discard
1 wood spacer measuring 1/8" x 1/2" x 2" (or a pencil)
1 marking pen

Materials required for mutant isolation:

1 24 hour, nutrient broth culture of *E. coli*
1 sterile 1 ml pipet
1 canister containing 95% isopropyl alcohol (or ethanol) for pipet discard
1 bent glass rod or spreader
1 beaker containing 95 % isopropyl alcohol (or ethanol)
1 bunsen burner and striker (or matches)
1 marking pen
1 inoculating loop

General Lab Equipment:

1 incubator set at 37°C
autoclave
waterbath set at 80°C for maintaining deeps if they are premelted

Pre-lab Preparation

Timetable of Events

Lab Period 1: 1.5 - 2 hours. Make and inoculate gradient plates.

Lab Period 2: 0.5 - 1 hour. Examine gradient plates and streak colonies.

Lab Period 3: 0.25 - 0.5 hours. Examine streaked colonies and record results.

Calculations are for each pair or group of students:

1. Prepare one nutrient broth culture of streptomycin-sensitive *E. coli*. The culture should be 24 hours old.

2. Prepare two nutrient agar deeps per pair. Deeps contain 10 ml of nutrient agar. It saves time to melt the deeps for students rather than to have them do the melting. Presterilized deeps can be melted by autoclaving for 10 minutes prior to class. Melted deeps can be maintained in an 80°C waterbath for several hours.

   Stock streptomycin solution. This solution should be made by adding 10 mg streptomycin per 100 ml sterilized, distilled water.

Method

Procedure for Gradient Plate Preparation

1. Melt two nutrient agar deeps and maintain them in liquid state.

2. Place the wood spacer under one edge of the empty, sterile petri plate.

3, Cool one of the melted deeps to approximately 45°C.

4. Pour the contents of the cooled, nutrient agar deep into the plate. Be sure that the medium covers the entire bottom surface of the plate. When the agar has solidified completely, remove the wood spacer. Move on to step 5 while you are waiting for the agar to solidify.

5. Cool the second melted deep to approximately 45°C.

6. Pipet 0.1 ml of the streptomycin solution into the second agar deep. Cap the deep and discard the pipet properly. Rotate the

Method, continued

> **Protocol Hints**
>
> 1. Do not pour over the sides of the plate
> 2. Do not forget to remove the wood spacer or forget to use the wood spacer

deep gently between the palms of your hands to mix the agar and the streptomycin solution.

7. When the agar in the petri dish has solidified, pour the contents of the second deep carefully over the top. Remember that the wood spacer should be removed before the second deep is poured.

8. When the contents of the second deep have solidified, label the high and low streptomycin concentration areas of the plate. Use a marking pen and write on the bottom of the plate.

Procedure for Mutant Isolation

1. Place the glass spreader in the beaker containing 95% isopropyl alcohol.

2. Using the aseptic technique, pipet 0.2 ml of the *E. coli* broth onto the surface of the gradient plate. Discard the pipet in the pipet canister.

3. Remove the glass spreader from the beaker and gently shake off the excess alcohol. While holding the spreader down and away from your body, quickly pass it through the flame of the bunsen burner. The alcohol residue will burn off quickly.

4. Use the glass spreader to spread the *E. coli* broth over the surface of the gradient plate. Spread the organism over the entire surface of the plate. Return the spreader to the beaker containing alcohol when you are finished.

5. Invert and incubate the plate at 37°C for 48 hours.

6. Observe the plates in the area of high streptomycin concentration for any streptomycin resistant mutants. Count the number of colonies in this area and record your results.

7. Select several well-isolated colonies in the mid to low concentration area of the plate. Using a sterile inoculating loop, streak each chosen colony over the surface of the medium toward the higher concentration area. Return the plate to the 37°C incubator for an additional two days.

8. Examine the plate again to note the effect of streaking the colonies. What effect does this have on their growth?

Results

First observation made after two days incubation - observe the plates in the area of high streptomycin concentration for any streptomycin resistant mutants. Count the number of colonies in this area and record your results.

Second observation made after two days of additional incubation - examine the plate again to note the effect of streaking the colonies. Record your observations.

Report Sheet

1. How many colonies were observed in the high concentration area of the plate?______________________________

2. Draw a picture of the plate. Label the high and low concentration areas. Illustrate the colony distribution on the plate.

3. What effect did streaking colonies to higher areas of concentration have on their growth?

____

____

____

4. Can you explain the effect described in question 3?

____

____

____

5. Make a second drawing of the plate and illustrate the effect of streaking the colonies into the high concentration area of the plate.

UNIT 4: MUTAGENESIS

MODULE 14: SPONTANEOUS MUTATION - INDIRECT SELECTION

Audrey M. Brown

* Introduction
* Safety Guidelines
* Materials
* Pre-lab Preparation
* Method
* Results

Introduction

The purpose of this unit is to illustrate how indirect (negative) selection is used to detect a spontaneously formed, streptomycin resistant mutant of *E. coli*. Replica plating is used to inoculate several types of test media.

A mutation is any change in the nucleotide sequence of DNA. These changes are permanent, genotypic changes. Although most mutations are harmful to the organism, a small percentage are beneficial under certain environmental conditions. Most mutations occur during chromosome duplication (cell division) and they may be spontaneous or they may be induced by mutagenic agents. Spontaneous mutations are thought to occur without any known cause. The rate of spontaneous mutation varies with each gene. In other words, some genes mutate at a faster rate than others. The rate of spontaneous mutation ranges from one in every thousand cell divisions (1/1000) to one in every billion cell divisions (1/1,000,000,000). The average rate is one in every million cell divisions (1/1,000,000). A mutagenic agent or mutagen is any substance that increases the rate above that occurring spontaneously. Examples include certain chemicals, mutagenic viruses, and radiation. In this experiment, we will determine whether the mutation is spontaneous or caused by the drug streptomycin. The streptomycin resistant mutants of *E. coli* will then be identified on streptomycin agar by the use of replica plating.

When bacteria are plated directly on a medium containing streptomycin, a very important question arises: Are resistant mutants occurring spontaneously or is the drug inducing the mutations? To answer this question, we will plate bacterial cells on nutrient agar and then test the colonies to see if they possess a mutation that confers streptomycin resistance. In this case, we may assume that streptomycin resistance is due to a spontaneous mutation since the bacteria have not been exposed to the drug prior to the test.

In order to test the bacteria growing on the nutrient agar plate for streptomycin resistance, it will be necessary to transfer many hundreds of colonies to the test medium (streptomycin agar). This task can be quite time consuming and laborious. For this reason we will employ a procedure called replica plating. In this procedure, sterilized velveteen is stretched over a replica plating stand. The nap or fibers of the velveteen act like hundreds of inoculating needles. When an entire plate, called the master plate, is placed on the velveteen surface, its colonies form an imprint that can be used to inoculate a series of test media. Test plates are aligned on the velveteen and inoculated one after another. Typically no more than four test plates are inoculated with one replica plating system. In this exercise we will inoculate one plate of nutrient agar and one plate of streptomycin agar. By examining the growth on these plates, we can determine whether the streptomycin-resistant colonies were present in the culture before exposure to the drug.

Safety Guidelines

Students should be trained in basic microbiology techniques including but not restricted to: aseptic techniques, spread plating, pipeting procedures and discarding, and proper handling and transfer of bacterial cultures.

Used plates, bacterial cultures, pipets and contaminated materials should be properly autoclaved before discarding in a waste container.

Materials

Materials have been calculated for one pair or group of students.

First lab period:
Making the Master Plates

1 nutrient broth culture of *E. coli*
1 nutrient agar plate
1 sterile 1 ml pipet
1 canister containing 95% isopropyl alcohol for pipet discard
1 bent glass rod or spreader
1 beaker containing 95% isopropyl alcohol
1 bunsen burner and striker (or matches)
1 marking pen

Second lab period:
Indirect (Negative) Selection Using Replica Plating

1 nutrient agar plate culture of *E. coli* from previous lab period
1 nutrient agar plate
1 nutrient agar plate containing 100 μg streptomycin per ml of medium called streptomycin agar
1 replica plating stand and clamp
1 piece sterilized velveteen in a glass petri plate
1 petri dish or discard container for used velveteen
1 pair of forceps soaking in a beaker containing 95% isopropyl alcohol
1 marking pen

Third lab period:
Analyzing the Replica Plates

1 master plate made during the first lab period
1 nutrient agar and one streptomycin agar replica plate made during the second lab period
1 Quebec colony counter and hand tally

Materials, continued

Velveteen is relatively expensive. It is possible to reuse the velveteen if it is autoclaved to kill bacteria and then washed and stored until the next time the lab is done. It will be necessary to autoclave the velveteen before it is used again.

Velveteen can be autoclaved in a glass petri plate that is wrapped with aluminum foil. The foil prevents steam from penetrating through the openings in the glass petri plate and therefore keeps the velveteen dry. Alternatively, the glass petri plates may be placed in a plate canister for autoclaving if one is available.

Pre-lab Preparation

Timetable of Events

First Lab Period : Only about 30 minutes are required to prepare master plates if students have been previously trained in pipeting and spread plating techniques. Demonstrate and discuss each of these techniques if they are new to the students.

Second Lab Period : Allow approximately 30 minutes for each student to analyze their master plates and fifteen minutes to demonstrate the replica plating technique. Students can then complete the replica plating in about 30 minutes.

Third Lab Period : Allow 30 minutes to one hour for students to analyze and discuss their test plates.

First Lab Period - Making the Master Plates:

Prepare one 48 hour nutrient broth culture of *E. coli* for each pair or group of students.

Prepare or purchase one nutrient agar plate per pair of students. Additional plates required during second lab period. It is easier to make one batch of media to supply both laboratory sessions.

Second Lab Period - Indirect Selection using Replica Plating:

Prepare one nutrient agar plate per pair or group of students.

Prepare one nutrient agar plate containing 100 μg streptomycin per ml of medium, called streptomycin agar, for each pair or group of students.

Third Lab Period - Analyzing the Replica Plates:

No special preparation is required for this session.

Method

First Lab Period: Making the Master Plates

1. Place the bent glass rod or spreader into the beaker containing 95% isopropyl alcohol.

2. Using aseptic technique, pipet 0.1 ml of *E. coli* from the broth culture to the surface of the nutrient agar plate.

3. Remove the glass spreader from the beaker and gently shake off the excess alcohol. While holding the spreader down and away from your body, quickly pass it through the flame of the bunsen burner. The alcohol residue will burn off quickly.

4. Use the glass spreader to spread the *E. coli* broth over the surface of the agar plate. Spread the organism over the entire surface of the plate. Return the spreader to the beaker containing alcohol when you are finished.

5. Incubate the plate in an inverted position (bottom side up) at 37°C for 24 to 48 hours.

Second Lab Period: Indirect Selection Using Replica Plating

1. Examine the *E. coli* plates made during the previous laboratory period. There should be a number of well-separated colonies on each plate. These plates are called master plates.

2. Use a marking pen to draw a small arrow on the bottom of your master plate. This arrow will serve as an alignment mark for replica plating, and later for comparing it to test plates.

3. Obtain one plate of nutrient agar and one plate of streptomycin agar. Label each plate. Use a marking pen to draw a small arrow on the bottom of each plate in the same way that you did for the master plate.

4. Prepare the replica plating device by removing the sterile velveteen from its petri plate with a pair of forceps. The forceps should be soaked in alcohol and flamed before use. Carefully place the velveteen over the replica plating stand and fasten it with the clamp provided. Your instructor will demonstrate this technique before you start. Do not touch the velveteen with your hands or contaminate its work surface.

5. Remove the lid from the master plate and place the plate over the velveteen surface so the alignment arrow is pointing away from your body. Gently press the plate down onto the velveteen with your hand. Do not move the plate around, just press it down from above.

Method, continued

6. Remove the master plate and replace its lid. Now remove the lid from the nutrient agar plate and place it over the velveteen surface so its alignment arrow is pointing away from your body (in the same direction as that used for the master plate). Press the plate gently into the velveteen surface.

7. Remove the plate from the velveteen and replace its lid.

8. Remove the lid from the streptomycin agar plate. Place the plate over the velveteen so its alignment arrow is pointing away from your body and gently press down on the bottom of the plate.

9. Remove the plate and replace its lid.

10. Remove the clamp from the replica plating device. Using forceps, remove the velveteen and discard it in the appropriate container.

11. Incubate the plates in an inverted position (bottom side up) at 37°C for 48 hours. Store the master plate at room temperature or in a standard refrigerator if one is available.

Third Lab Period: Analyzing the Replica Plates

1. Examine the replica plates and compare them to the original master plate. Count the number of colonies on each of your test plates. Complete the analysis described in the results section of this form.

Results

1. Examine the replica plates and compare them to the original master plate. Does the nutrient agar plate look exactly like the original master? Examine the streptomycin resistant mutants growing on the streptomycin plate. Can you locate these colonies on the original master and the nutrient agar plates?

2. Count the number of streptomycin resistant mutants growing on the streptomycin agar plate. Do a total plate count for the number of colonies growing on the nutrient agar replica plate.

Report Sheet

Student Group Number	**Number of Colonies Counted**	
	Nutrient Agar	**Streptomycin Agar**
1		
2		
3		
4		
5		
6		
Class Average		

UNIT 4: MUTAGENESIS

MODULE 15: INDUCED MUTATION

Audrey M. Brown

Introduction

The purpose of this experiment is to analyze the mutagenic effects of ultraviolet radiation on *Serratia marcescens.*

A mutation occurs any time there is a change in the nucleotide sequence of DNA. These changes are permanent. Spontaneous mutations are thought to occur without any known cause. A mutagenic agent or mutagen is any substance that increases the rate above that occurring spontaneously. Certain chemicals, mutagenic viruses and radiation are known as mutagenic agents. In this experiment, the mutagenic effects of ultraviolet light are analyzed in a pigment producing strain of *Serratia marcescens.*

Serratia marcescens is a gram negative, rod shaped bacterium that is classified along with *E. coli* in the enteric group. When *Serratia* is grown at room temperature it forms red-colored colonies on nutrient agar. Pigment production is inhibited by high temperatures and therefore when it is grown at 37°C, it often fails for form its red pigment. This type of variation is temporary and is characterized by any change in behavior or appearance brought about by environmental conditions. In this experiment, we will attempt to cause a mutation that inhibits pigment production under any environmental condition (including growth at room temperature). This change is permanent because the gene for pigment production is altered and therefore rendered nonfunctional.

The mutagenic agent used in this experiment is ultraviolet radiation. This is the same type of radiation that causes tanning in humans. UV radiation causes structural changes in DNA called dimers. Many bacteria possess repair mechanisms that repair dimer-damaged DNA. One such mechanism is called the light repair process or photoreactivation. In this repair process a photoreactivating enzyme forms a complex with the dimer. If visible light is present it provides the energy necessary for this enzyme to break the dimer bond and correct the structure of DNA. To avoid this repair mechanism and allow for the maximum number of mutants, we will incubate irradiated bacteria in the dark.

Safety Guidelines

Students should be trained in basic microbiology techniques including but not restricted to: aseptic techniques, spread plating, pipeting procedures and discarding, and proper handling and transfer of bacterial cultures.

Used plates, bacterial cultures, swabs and contaminated materials should be properly autoclaved before discarding in a waste container. Be sure that students use the UV lamp properly and that they wear protective glasses.

Materials

Materials have been calculated for one pair or group of students.

1 nutrient broth culture of *Serratia marcescens*
5 nutrient agar plates
1 pair ultraviolet radiation protective glasses/goggles per student
1 ultraviolet lamp (germicidal quality)
5 sterile swabs (cotton tipped applicators)
1 swab discard container with 95% isopropyl alcohol
1 marking pen
1 incubator set at 25°C or a room temperature incubation area
1 petri plate canister or box to allow for incubation in a dark environment
autoclave

Pre-lab Preparation

Lab Period 1: 1.5 - 2 hours. Make and inoculate gradient plates.

Lab Period 2: 0.5 - 1 hour. Examine gradient plates and streak colonies.

Lab Period 3: 0.25 - 0.5 hours. Examine streaked colonies and record results.

Prepare one 24 to 48 hour nutrient broth culture of *Serratia marcescens* per pair or group of students. The strain should produce a red colored pigment when grown at room temperature.

Prepare 5 nutrient agar plates per pair of students.

Method

1. Obtain five nutrient agar plates. Label the bottom of the first plate "2 seconds". Label the bottom of the second plate "5 seconds". Label the third plate "10 seconds", the fourth plate "20 seconds" and the fifth plate is labeled "control".

2. Suspend broth culture by gently tapping culture tube against the palm of your hand. Do not shake or spill the culture.

3, Dip one sterile swab into broth culture using aseptic technique. Prepare lawn of bacteria on plate surface by completing the following directions. Inoculate the surface of the first plate by moving the swab carefully over its entire surface. Rotate the plate and the swab, and cover the plate again in a second direction. Rotate the plate and cover the surface of the agar in a third direction and then swab around the perimeter of the plate. Discard the swab in the alcohol discard container.

4. Inoculate each of the four remaining plates with separate swabs using the technique described in step 3.

5. Be sure to wear UV protective glasses before proceeding. Allow ultraviolet lamp to warm up a few minutes. The wavelength of UV radiation should be approximately 265 nm and the light should be about 6 inches from plates when used.

6. Remove the cover of the first plate and expose the lawn of bacteria for the time designated on the label (2, 5, 10 or 20 seconds). Replace the cover quickly. Expose the remaining plates for their designated times. Note that the control plate is not exposed to the UV radiation.

7. Incubate the plates in an inverted position at 25°C (or at room temperature) in a dark environment for 3 to 5 days. A dark environment such as a petri plate canister or box is essential. Visible light may stimulate the photoreactivation repair mechanism present in many bacteria and therefore remove the mutations. The largest source of errors in this experiment is from incubating colonies under visible light. Incubation in the dark is essential.

8. Observe the plates for the appearance of nonpigmented (white) mutants. Compare these with the typical, pigmented (red) colonies on the plate and with those on the control plate. Note how the total number of colonies on the plates decreases as the exposure time increases.

9. Record your results in the Results section. Indicate the number of typical, red colonies and the number of mutant, white colonies for each exposure time. Can you draw any conclusions from this?

Results

1. Observe the plates for the appearance of nonpigmented (white) mutants. Compare these with the typical, pigmented (red) colonies on the plate and with those on the control plate. Note how the total number of colonies on the plates decreases as the exposure time increases.

2. Record your results in the Results section. Indicate the number of typical, red colonies and the number of mutant, white colonies for each exposure time. Can you draw any conclusions from this?

Report Sheet

Exposure Time To UV Light	Number of Colonies of *Serratia marcescens*	
	Pigmented (Red)	Nonpigmented (White
2 seconds		
5 seconds		
10 seconds		
20 seconds		
Control (0 seconds)		

UNIT 4: MUTAGENESIS

MODULE 16: *IN VITRO* MUTAGENESIS

Bruce Collier

* Introduction
* Safety Guidelines
* Materials
* Pre-lab Preparations
* Method
* Results

Introduction

Recently, one of the most powerful tools to be added to the molecular biologist's toolchest is a technique called *"in vitro* mutagenesis." Prior to the application of this technology, molecular biologists were limited to "cutting and pasting" pieces of DNA together to accomplish specific DNA sequence changes. If they wanted to change DNA sequences they were limited to trying to use clever restriction endonuclease patching tricks or synthesizing synthetic oligodeoxyribonucleotides. There were chemical mutagenesis techniques but these did not give the specific and desired changes to the DNA sequence.

Then *in vitro* mutagenesis was developed and refined to the point which allowed a specific DNA modification. While there are now a few minor variations on this approach, the first approach is still the basis for this technique and will be focused on in the rest of this discussion. The technique will be referred to as "Kunkel's *in vitro* mutagenesis" or just "*in vitro* mutagenesis."

Using Kunkel's *in vitro* mutagenesis, molecular biologists can change one or a few DNA sequences in a piece of DNA specifically. They can do this to put a new restriction endonuclease recognition site into their piece of DNA, or change the DNA sequence of a triplet codon DNA sequence to optimize expression, or change the triplet codon entirely to put a new amino acid into the protein to study the effect of the new amino acid on protein function.

Now that we know some of the uses of Kunkel's *in vitro* mutagenesis, we should find out the technology of how it is accomplished. In order to understand how this system works we need to understand DNA polymerase, M13 bacteriophage biology, *E. coli* dut and ung gene mutants, DNA hybridization, synthetic oligonucleotides, kinase and ligase enzymes.

DNA polymerase is the enzyme that allows living cells to replicate their own DNA by making a new copy of DNA that then gets placed into a new cell. DNA polymerase requires a template, which must be DNA. This is the piece of DNA that it uses to generate a copy of itself. It also needs a primer. This primer is a short piece of DNA that is hybridized to the template and is the place where DNA polymerase starts its elongation process (as in Figure 1). The primer and the template hybridize and line-up with each other according to Watson-Crick rules (A:T and C:G).

Introduction, continued

```
   5'-CGTCGTGACTGGGAAAACCCT-3'
      |||||||||||||||||||||
3'-TTGCAGCACTGACCCTTTTGGGACCGCAATGGGTTGAATTA-5'
```

Figure 1: A Primer Hybridized to a Template

The primer must have a 3'-hydroxyl group to allow the addition of the next nucleotide in the sequence, based on the template complementary sequence. Molecular biologists often refer to the 5' (5 prime) to 3' (3 prime) direction that polymerase moves in. One can easily remember this if they recall that the DNA primer must have a 3'-hydroxyl group and why it is there. Therefore, you can have DNA polymerase synthesize a new DNA strand from wherever a DNA primer is hybridized to the template strand. This phenomenon is utilized in one of the popular DNA amplification strategies called polymerase chain reaction or PCR for short.

In order to generate a new strand, DNA polymerase also requires the deoxyribonucleotide that is complementary to the template nucleotide. Again this follows the Watson-Crick rules of pairing. In the figure above, a 'G' or guanine deoxyribonucleotide is required for the next nucleotide that will be incorporated. DNA polymerase uses the deoxyribonucleotides in the form of deoxyribonucleotide triphosphates. The triphosphate form helps to supply the energy required for the DNA polymerization reaction. DNA polymerase also uses magnesium as a cofactor.

DNA polymerases have a proofreading or editing function that helps to reduce the errors in copying from the template. This function is not extremely important in order to understand *in vitro* mutagenesis, but is included to demonstrate one of the drawbacks of this technology. DNA polymerases usually only generate mutations once in every 10 to 100 thousand nucleotides so it is not a serious problem. However, molecular biologists usually sequence their DNA once they have performed *in vitro* mutagenesis to be sure they have not generated a mutation in their DNA sequence some other place, other than the region where they wanted a mutation.

Finally, different DNA polymerases may exhibit a phenomenon called "strand displacement." Imagine a circular single stranded DNA plasmid molecule with a DNA primer annealed to it like Figure 2A.

Introduction, continued

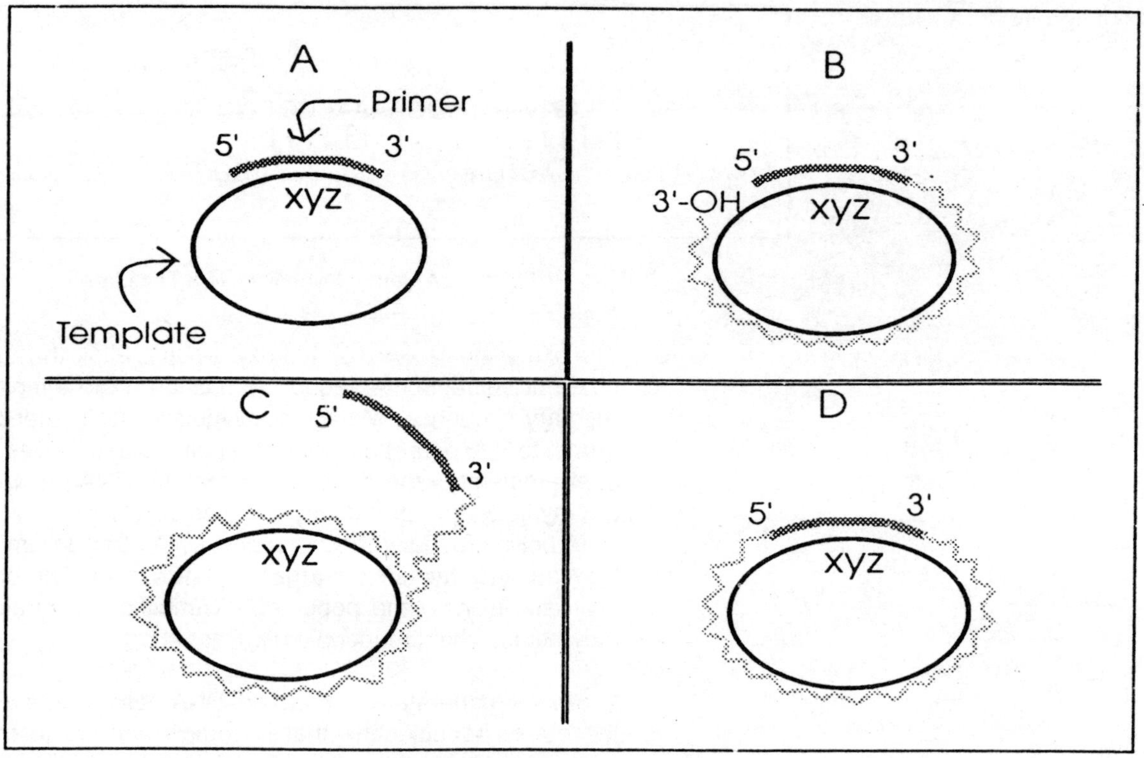

Figure 2: Strand Displacement

The DNA polymerase will make a DNA copy of the template DNA, starting from the DNA primer moving in the 5' to 3' direction. What happens when the DNA polymerase gets back around to the 5' end of the hybridized primer molecule as shown in Figure 2B?

If the DNA polymerase can perform "strand displacement", it will push the original primer out of the way and continue to elongate around the plasmid template, as shown in Figure 2C. This can actually occur several times, generating a copy of the DNA that is several times the size of the original plasmid.

For Kunkel mutagenesis we want the DNA polymerase to stop when it gets back to the original DNA primer, therefore we prefer a DNA polymerase that lacks the ability to perform strand displacement. A DNA polymerase from a virus of *E. coli*, bacteriophage T4 has such a DNA polymerase enzyme, which is referred to as T4 DNA polymerase.

Imagine we now have a single stranded circular plasmid template that has a primer hybridized to it, and T4 DNA polymerase has synthesized a new strand all the way around and stopped at the 5' end of the primer as in Figure 2B.

Introduction, continued

DNA ligase joins two pieces of DNA. One with a 3'-hydroxyl, the other with a 5'- phosphate. They must be side by side. DNA ligase then forms a phosphodiester bond with energy supplied by adenosine triphosphate (ATP).

You will notice that the newly synthesized strand is not completely joined together, it still has a "nick." In order to join the 3' end of the newly synthesized strand to the 5' end of the primer we need another enzyme called ligase. Ligase is a DNA joining enzyme. Ligase generally only works on double stranded DNA and requires a 3'-hydroxyl and 5'-phosphate group on two different DNA pieces at a nick site to be joined in order to perform the joining. When using synthetic oligonucleotides it is important to make sure that they have a 5'-phosphate group, before beginning the experiment. This can be added chemically during the synthesis or it can be added later using an enzyme called kinase. After using ligase on this newly synthesized molecule we now have a double stranded circular molecule (see Figure 2D).

The next key to this technique is the generation of the original single stranded DNA that can be used as a template for the DNA polymerization reaction. To accomplish this we exploit a different virus of *E. coli* called bacteriophage M13. This virus exists as both a double stranded and single stranded DNA form. The gene to be mutagenized is cloned into the double stranded form of the virus using standard restriction endonuclease techniques. This recombinant double stranded DNA is then transformed into competent *E. coli* cells. Once transformed the DNA will be replicated in a double stranded DNA form inside the cell just like a plasmid. A unique feature of this bacteriophage is that it can also replicate as a single stranded DNA molecule form. All the single stranded DNA is only one specific DNA strand, called the plus (+) strand. This plus strand gets packaged into a bacteriophage coat protein, a product of one of the bacteriophage genes. These coat proteins are proteins that protect the DNA when it is outside of the *E. coli* cell. Secreted bacteriophage particles can then infect uninfected *E. coli* cells until all cells are infected. In order to infect the uninfected cells the bacteriophage must attach to F-pilus (as shown in Figure 3).

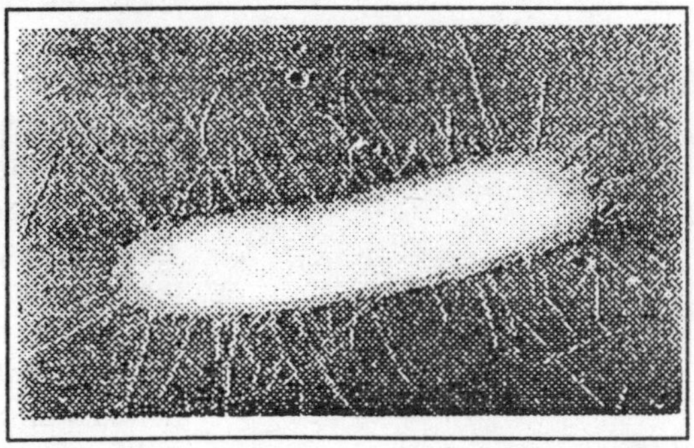

Figure 3: *E. coli* with Pili

Introduction, continued

Not all *E. coli* cells generate F-pili structures. F-pili are made by genes that are found on special plasmids called F factors or F episomes. These F factors are only present in a few copies in the cell. It is important when growing these bacteriophage that the *E. coli* cells have F factors that can make the F-pili. If present, these F factor containing cells can be maintained by antibiotic selection.

In order to make single stranded DNA the infected *E. coli* cells are grown in liquid culture then the *E. coli* cells are removed and pelleted by centrifugation. The bacteriophage are left in the supernatant. This effectively purifies bacteriophage single stranded DNA from *E. coli* chromosomal and F-factor and any other DNA still inside the cell. Using polyethylene glycol and sodium chloride the bacteriophage can be harvested by centrifugation. The protein coats on the bacteriophage can be removed by phenol and detergent extraction, leaving reasonably pure single stranded DNA.

```
    GlnArgArgAspTrpGluAsnProGly            wild type
              *                         protein and DNA
5'-CAACGTCGTGACTGGGAAAACCCTGGC-3'          sequence
3'-GTTGCTGCTCTGACCCTTTTGGGACCG-5'

    GlnArgArgAspAmbGluAsnProGly             mutant
              *                         protein and DNA
5'-CAACGTCGTGACTAGGAAAACCCTGGC-3'          sequence
3'-GTTGCAGCACTGAGCCTTTTGGGACCG-5'
```

Figure 4: Sequence of part of the wild type and mutant β-galactosidase

Now we have a method to generate single stranded DNA template, but how do we generate a specific mutation from this template? We know that DNA polymerase requires a short DNA primer as shown in Figure 1.

In Figure 4 we want to convert the codon that codes for Tryptophan in the β-galactosidase structural gene to a codon that codes for an amber or "stop translation" codon. The nucleotide that is different is indicated by an asterisk. Therefore, one nucleotide in the mutagenic primer sequence compared to the natural primer is different. The mutagenic primer still maintains a large degree of complementarity with the template and can hybridize to the template as shown in Figure 5. The mutagenic primer is used by DNA polymerase to generate a new DNA strand from the natural (wild type) template.

Introduction, continued

```
5'-CAACGTCGTGACTGGGAAAACCCTGGCGTTACC-3'  secreted template
    ||||||||||| |||||||||||
   3'-GCAGCACTGA CCTTTTGGGAC-5'      mutagenic primer
               T
```

Figure 5: Hybridization of Mutagenic Primer to the Natural DNA Sequence

In this experiment we can also check that the mutation is due to the amber mutation caused by the *in vitro* mutagenesis and not due to some other mutation. Normally, the enzyme β-galactosidase is present and cleaves lactose to form galactose and glucose. We use an analogue of lactose, XGAL (5-bromo-4-chloro-3-indolyl-β-D-galactopyranoside) which turns blue when cleaved by β-galactosidase. The amber mutation of β-galactosidase will not make a functional enzyme and the blue color will not be generated.

We can also determine if the nonfunctional β-galactosidase protein is specifically due to the amber mutation. There are specific strains of *E. coli* that contain suppressor tRNA genes. Specific suppressor tRNAs recognize the amber mutation 5'-UAG-3' and places a glutamine amino acid at this position, allowing translation to continue. This allows a functional β-galactosidase gene to be synthesized in these suppressor strains of *E. coli.* Therefore, these suppressor strains will produce a blue color with XGAL.

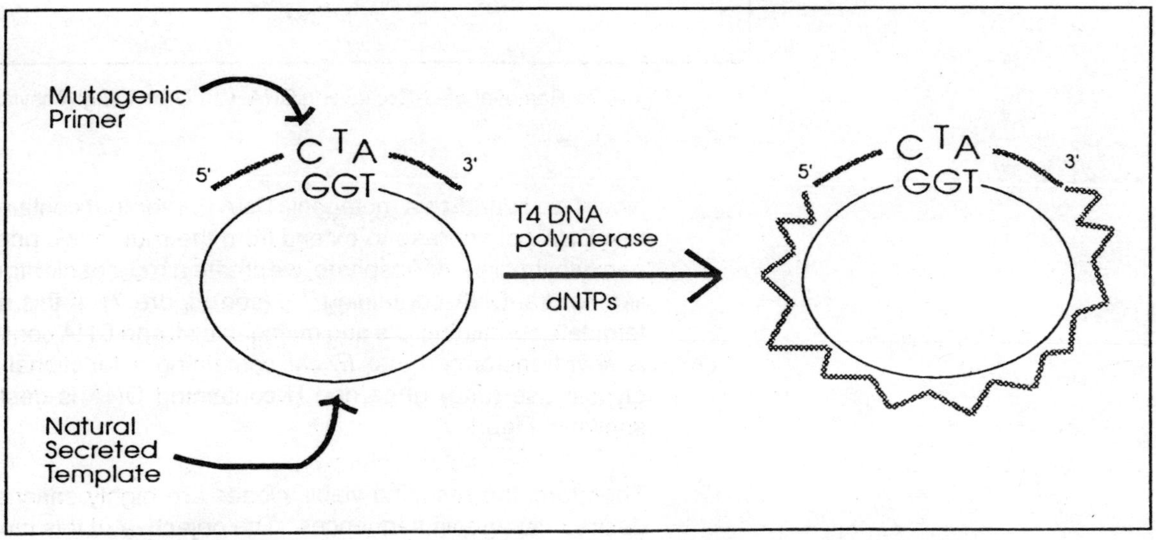

Figure 6: Incorporation of Mutagenic Primer in Newly Synthesized DNA Strand

Introduction, continued

Once the mutagenic primer is incorporated into the new strand by extending the sequence on the template with DNA polymerase (see Figure 6), two different kinds of DNA strands are present. The original natural strand and the newly synthesized mutagenized strand are present. Since both strands can replicate after transformation, molecular biologists used another molecular trick to select specifically for the mutagenized DNA strand. If they use the *E. coli* genes dut and ung they can accomplish this feat. Dut and ung are the gene mnemonics that stands for dUTP pyrophosphatase and Uracil N-glycosylase, respectively. By using an *E. coli* strain that is mutant or defective for both of these genes single stranded DNA can be made which contains Uracil (U) in place of the naturally occurring thymine (T). Therefore, single stranded DNA from such an *E. coli* strain contains U instead of T.

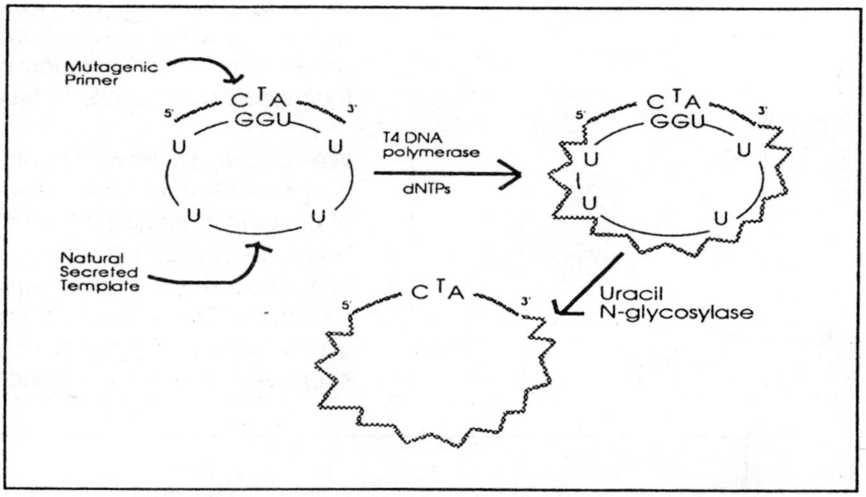

Figure 7: Removal of U-Containing DNA with Uracil-N-glycosylase

Now if we hybridize a mutagenic DNA primer that contains T's and use DNA polymerase to extend from the mutagenic primer using deoxythymidine triphosphate, we create a mutagenic strand that is like natural DNA containing T's (see Figure 7). If this mixture of template containing U's and mutagenic strand DNA containing T's is now transformed into *E. coli* containing a functional Uracil N-glycosylase (ung) gene, the U containing DNA is destroyed as shown in Figure 7.

Therefore, the resulting viable clones are highly enriched for the desired mutagenic sequences. The objective of this module is to study the use of in vitro mutagenesis to make specific DNA sequence changes and cause structural gene changes.

Safety Guidelines

Safety glasses and gloves should be worn when performing this module. Biological and other material should be disposed of in specially labelled containers. When using hot agar and agarose be careful of hot samples, which can inflict severe burns.

Biological material should be disposed by autoclaving or soaking the material in a 1% bleach solution overnight.

Materials

This module is based on EDVOTEK Experiment #304. The experiment includes the following components:

A. Uracil containing phage DNA
B. Mutagenic oligo primer
C. 10X annealing buffer
D. T4 DNA polymerase
E. T4 DNA ligase
F. 10X extension buffer
G. Sterile water
H. Control transformation DNA
I. JM109 competent cells
J. JM109 cell slant
K. Suppressor cell slant
L. XGAL/IPTG plates
M. Recovery broth
N. Stop buffer
O. Bacteriophage controls

Requirements

1-20 μl and 10-100 μl micropipetor and sterile pipet tips for each group

37°C water bath

42°C water bath (since this is used for a short period of time the water can be heated, then cooled with ice prior to that experimental step)

65°C water bath

Sterile microfuge tubes

Ice (preferably crushed)

Materials, continued

37˚C incubator

37°C shaker

JM109 and suppressor cells (grown by instructor)

5 and/or 10 ml sterile pipets

Low speed tabletop centrifuge (or preferably a microcentrifuge)

-20°C freezer

Microwave or hot plate to prepare top agarose and agar for plates

100 ml beaker

Floats for water baths

Tubes for growing phage cultures

Pre-lab Preparation

Time Requirements: A 4 day lab, requiring 6 hours
2 hour maximum time requirements

Dispense Equipment

- 1-20 µl pipetors and sterile pipet tips
- 1.5 ml sterile microcentrifuge tubes
- T4 DNA polymerase (on ice)
- T4 DNA ligase (on ice)
- 10X annealing buffer (on ice)
- 10X extension buffer (on ice)
- ice and ice buckets
- 65°C water bath
- 37°C water bath
- 100 ml of sterile water
- mutagenic primer (about 5 pmol) (on ice)
- uracil containing template DNA (about 0.1 pmol) (on ice)
- 100 or 250 ml beakers
- floats for tubes in water baths
- stop buffer
- -20°C freezer for overnight storage

Pre-lab Preparation, continued

Dispense Equipment and Competent Cells

- freshly thaw JM109 competent cells
- aliquot 1 ml of control DNA
- 42°C water bath
- 37°C shaking water bath or air incubator
- 2 XGAL/IPTG plates per group
- 37˚C incubator
- recovery broth (about 10 ml)
- 1 sterile 1, 5 or 10 ml pipet

Dispense Sterile Plastic Ware and Media

- 5 sterile culture tubes
- aliquot recovery broth (about 30 ml) per group
- aliquot freshly grown JM109 cells (about 1 ml)
- 1 sterile 5 or 10 ml pipet
- 1 sterile 1 ml pipet
- 1-20 or 100 ml pipetor and sterile pipet tips
- 37°C shaking water bath or air incubator

Prepare Centrifuge and Dispense Equipment

- need centrifuge or preferably microcentrifuge
- 1 5 ml pipet
- 1-20 and 100 µl pipetor and sterile pipet tips
- 65°C water bath
- 2 XGAL/IPTG plates
- 42°C water bath
- prepared top agarose
- 1 ml each of overnight cultures of JM109 and suppressor
- (+) and (-) control phage preps
- 37°C incubator

Next Day Activities

Prepare Water Baths

- Pre-calibrate 37°C, 42°C, 65°C water baths
- Pre-calibrate 37°C water shaker or air shaker incubator

Prepare Plates and Streak out Cells

- Pour XGAL/IPTG plates
- When cooled streak out JM109 and suppressor cells

Overnight Culture of JM109

- Prepare overnight culture of JM109 cells from the streaks from B-2 (about 20 ml)

Overnight Culture of JM109 and Suppressor cells

- Prepare overnight culture of JM109 and suppressor cells from the streaks from B-2 (about 20 ml)

Method

Day 1

1. To a sterile microfuge tube add 4 µl of sterile water.

2. To this solution add:
   1 µl of 10X annealing buffer
   1 µl of mutagenic primer
   1 µl of uracil containing template DNA

3. Close the top of the tube and place it in a 65°C water bath for 5 minutes.

4. Remove about 100 µl of this hot water to a beaker and transfer the tube into this setup. This is to allow the annealing DNA to slow cool to room temperature. Leave the annealing DNA to cool for 30 minutes.

5. Cool the tube on ice for 5 minutes.

6. Centrifuge the microfuge tube in a microcentrifuge for 1 s.

7. To the tube add: (while sitting on ice): 1 µl of extension buffer, 1 µl of T4 DNA polymerase, and 1 µl of T4 DNA ligase

8. Close the top of the tube and let the tube sit on the bench for 5 minutes (ONLY).

9. Quickly place the tube in a 37°C water bath and incubate for 90 minutes.

10. Add 1 µl of stop buffer and place the sample at -20°C .

Day 2

11. Thaw DNA and add 100 µl of freshly thawed competent JM109 cells to the microcentrifuge tube. You will also need to set up control tubes, containing 1 µl of transformation control DNA and 1 µl of uracil containing template DNA. To these add 100 µl of freshly thawed competent JM109 cells. Incubate the cells on ice for 20 minutes.

12. Heat shock the cells for 90 s at 42°C.

13. Add 1 ml of recovery broth to the cells and shake the cells at 37°C for 60 minutes.

14. Spread the transformation mix onto an XGAL/ITPG plate.

15. Incubate the plate at 37°C inverted overnight. (The plates will have to be taken out of the incubator the next day.)

Method, continued

Day 3

16. Record data from experiment. The results table provided at the end of the Method section is to determine the ratio of white to blue bacteriophage plaques. The uracil containing DNA should NOT generate transformants in a dut+ ung+ host, therefore if there are a substantial number of transformants a mix up in the host cells may have occurred.

    The DNA control should generate many transformants. Occasionally some of these will not form blue plaques (~1%) due to the loss of the F episome which carries the lac I repressor.

    The experimental DNA should show over 80% white plaques, which have been *in vitro* mutagenized. If the ratio is lower it might suggest that the uracil incorporated DNA template was inadequately uracil substituted.

17. Set up 5 culture tubes with 5 ml of recovery broth and 0.1 ml of a freshly grown culture of JM109 cells. Label the tubes 1 to 5 and place your lab designation on the tube.

18. Use a sterile pipet tip on a pipetor to inoculate the 5 culture tubes. Take the sterile pipet and stab into a white phage plaque attempting to get the phage plaque in the pipet tip. Then mix culture tube contents with the inside of the pipet tip by drawing liquid into the pipet tip and releasing it several times. For number 5 pick a blue phage plaque if there is one.

19. Place the culture tubes to shake overnight at 37°C.

Day 4

20. Centrifuge and pellet *E. coli* cells. Remove 1.5 ml of overnight culture and transfer to a 1.5 ml microcentrifuge tube. Centrifuge the tube in a microcentrifuge tube for 15 minutes.

21. Carefully remove 100 µl of the supernatant to a clean sterile microcentrifuge tube.

22. Place the tube in a 65°C water bath for 5 minutes.

23. Prepare two plates with cells in the top agarose. Take 0.1 ml of freshly grown JM109 cells and transfer it to a culture tube. Add 3 ml of top agarose that has been liquefied then cooled to 42°C. Quickly transfer the agarose cell mixture to an XGAL/IPTG plate. Repeat this procedure, but use the suppressor cell line. Be sure to label the plates JM109 and suppressor.

Method, continued

24. Once plates have cooled and solidified, label plates with 7 spots spaced apart. Make sure to label two plates identically. Label the plates with 1 to 5 and positive (+) and negative.(-).

25. Spot 5 ml of each bacteriophage supernatant preparation onto spot positions 1 to 5 as appropriate. Also add 5 ml of (+) and (-) bacteriophage control phage preparation.

26. Do not disturb the plates while the spots are drying, allow about 10 minutes.

27. Incubate the plates at 37°C inverted overnight. Be sure to arrange for the plates to be removed the next day and record the data from the plates in the results table provided at the end of the Methods section. The experimental phage preparation should generally show a white color when grown on JM109 cells, and a blue color when grown on the suppressor cells. The mutation should NOT affect phage growth.

    The +ve control should be blue on both JM109 and suppressor cells, because this contains a wild-type β-galactosidase.

    The negative control should resemble the student data since it contains an amber mutation in β-galactosidase.

Method, continued

Result Table for Step 16

		DNA	
	Uracil	**Control**	**Experimental**
# of white phage plaques	____	____	____
# of blue phage plaques	____	____	____
Total number of plaques	____	____	____
Mutagenesis Ratio	____	____	____

$$\text{Mutagenesis Ratio} = \frac{\text{Number of White Plaques}}{\text{Total Number of Plaques}} \times 100 = \_\_\_ \%$$

Result Table for Step 27

Spot	**Growth of Phage (color)**	
	JM109	**suppressor**
Phage 1	____	____
Phage 2	____	____
Phage 3	____	____
Phage 4	____	____
Phage 5	____	____
(+)	____	____
(-)	____	____

indicate a + or - dependent on whether there was growth
indicate a (B) or (W) for color

Results

1. Observe the plates for the appearance of nonpigmented (white) mutants. Compare these with the typical, pigmented (red) colonies on the plate and with those on the control plate. Note how the total number of colonies on the plates decreases as the exposure time increases.

2. Record your results in the results table provided. Indicate the number of typical, red colonies and the number of mutant, white colonies for each exposure time. Can you draw any conclusions from this?

Results, continued

Sample Result Table for Step 16

	DNA		
	Uracil	**Control**	**Experimental**
# of white phage plaques	0	10	98
# of blue phage plaques	0	990	2
Total number of plaques	0	1000	100
Mutagenesis Ratio	0	1	98

$$\text{Mutagenesis Ratio} = \frac{\text{Number of white plaques x100}}{\text{Total Number of plaques}} = \underline{\ ?\ }\ \%$$

Sample Result Table for Step 27

Spot	**Growth of phage (color)**	
	JM109	**suppressor**
Phage 1	+(W)	+(B)
Phage 2	+(W)	+(B)
Phage 3	+(W)	+(B)
Phage 4	+(W)	+(B)
Phage 5	+(W)	+(B)
(+)	+(B)	+(B)
(-)	+(W)	+(B)

indicate a + or - dependent on whether there was growth
indicate a (B) or (W) for color

UNIT 5

Advanced Protocols for DNA Analysis

UNIT 5: ADVANCED PROTOCOLS FOR DNA ANALYSIS

MODULE 17: DNA SEQUENCING

Harley Mortensen

* Introduction
* Experimental Outline
* Safety Guidelines
* Materials
* Method
* Result

Introduction

The objective of this module is to perform DNA sequencing by the Sanger dideoxy method which includes sequencing reactions, pouring sequencing gels, electrophoresis, gel manipulation, and autoradiography. The Maxam-Gilbert method is an alternative sequencing procedure. Acquiring knowledge of nucleic acid sequences is at the heart of molecular biology. It reveals the genetic fine structure which has allowed the explosion in molecular biology theory to develop. An astounding amount of information has accumulated in a very short time about gene structure, regulation, mutational events, and so on; all due to the ability to determine sequences.

Dideoxy DNA sequencing is based on the principle that DNA polymerization (replication) on a template strand can be interrupted by the presence of dideoxyribonucleoside triphosphates (ddNTP) in the reaction mixture (Figure 1).

During polymerization, at any interval in time, a broad distribution of newly synthesized fragment sizes will exist. As chain extension of each of these fragments proceeds, a competition for the termini occurs when deoxy- and dideoxyribonucleotides (dNTPs and ddNTPs) are both present in the reaction mixture. ddNTPs will be occasionally incorporated into some of the growing DNA fragments at each position causing termination of the polymerization reaction for those fragments. dNTPs will be incorporated into the remaining growing fragments and will permit continued chain extension. Under properly adjusted conditions, a product mix can be generated that consists of all possible fragment lengths, each fragment being terminated by a ddNTP. The various fragment species can be separated from one another on a polyacrylamide gel to form a "ladder" in which each "step" differs in size from its vertical neighbors by a single nucleotide. If the fragments are labeled in such a way that they can be detected and read, this provides the sequence that is complementary to the template.

Once the sequence of one strand has been determined, the other strand sequence can be inferred by base pair complementarity. For technical reasons, however, uncertainties occur when sequencing. Therefore, each single strand is sequenced as a check against the accuracy of the sequence determined for the other.

Several vector systems have been designed which have been of the upmost importance in facilitating the development of sequencing technology. A series of M13 derivatives constitutes one family which takes advantage of the relative ease of preparation of SS DNA from viral vectors.

Introduction, continued

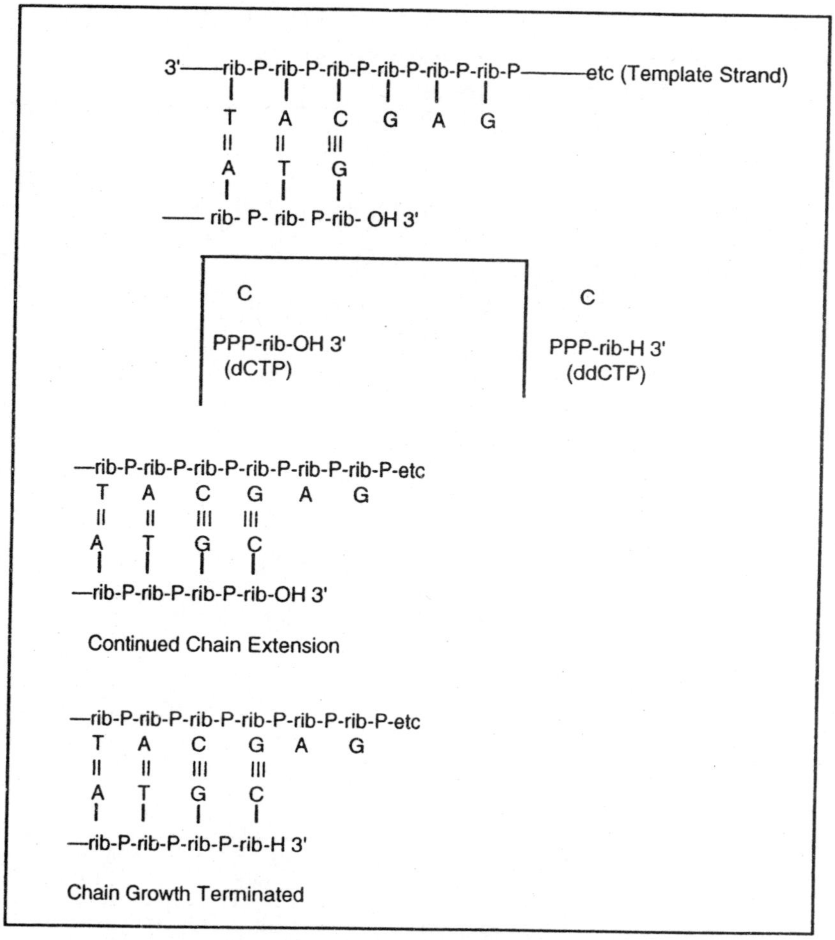

Figure 1: Dideoxy vs Deoxy Chain Termini
(All polymers are deoxy)

M13 Phage Vectors

The important characteristic of this system is that the DNA produced is single stranded and is extruded into the medium without lysis of the host. In this system, as opposed to purely plasmid vectors, the direct production of ssDNA simplifies the work because there is no necessity to separate complementary strands prior to sequencing. Use of ssDNA (single stranded DNA) allows the sequence of one strand to be determined without having to deal with difficulties caused by the presence of the complementary strand. ssDNA can be conveniently harvested by centrifugation of cellular material. The supernatant containing the ssDNA is phenol extracted to remove viral proteins followed by precipitation of the ssDNA with polyethylene glycol (PEG). Separation of the DNA from the PEG affords single stranded DNA for sequencing.

Introduction, continued

M13 has been modified to produce the vectors used in sequencing (Figure 2).

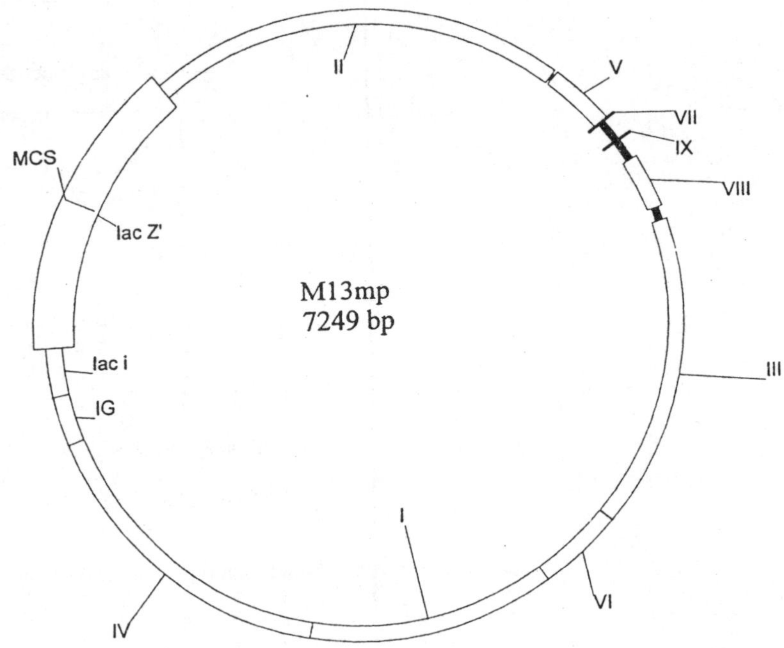

Figure 2: M13mp Map

The pioneer of this work is J. Messing. The vectors were engineered to contain a MCS (multiple cloning site), also referred to as a PCS (poly cloning site). The presence of several restriction sites within the MCS allows for a number of options for inserting foreign DNA in that region. Cutting at two different vector sites allows for "forced cloning"-DNA inserting in only one orientation- of foreign DNA provided the foreign DNA has been cut with the same two restriction enzymes. Another feature of the M13 vectors is their availability as pairs, each pair having the same MCS but in reverse orientation. With forced cloning, one strand will be produced by one member of the pair while the other member produces the complementary strand. Thus both strands are made available for sequencing. Different pairs have different MCSs.

Introduction, continued

An important aspect of the technology is the commercial availability of primers, which are oligonucleotides complementary to a sequence close to the MCS. When the commercially available primer is annealed to its complementary site on the vector, polymerization can be initiated from the primer in a controlled manner to produce the series of chain terminated fragments which ultimately become the sequence ladder upon electrophoresis.

Phagemids

Figure 3 shows the map of a pUC type phagemid vector.

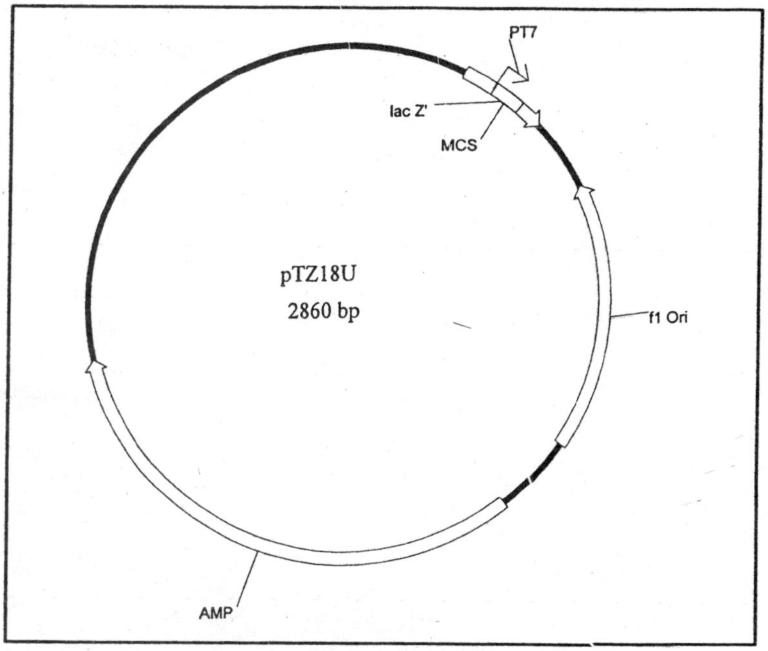

Figure 3: A Phagemid Map

In Figure 3, the plasmid origin of replication directs double strand replication as a plasmid. In the presence of a helper phage such as M13KO7, single strand replication occurs through the f1 origin and can be treated in the same manner as M13 single strand product. Alternatively, procedures have been developed for sequencing double stranded DNA produced from plasmid type replication. These vectors contain the same MCS as is found in M13 vectors so that operations such as subcloning are carried out in exactly the same manner. The same set of M13 primers can also be used when sequencing inserts within the MCS. As an additional feature, specific promoters (T7 promoter in Figure 2) are placed near the MCS allowing control of transcription through the insert DNA to produce RNA.

Experimental Outline

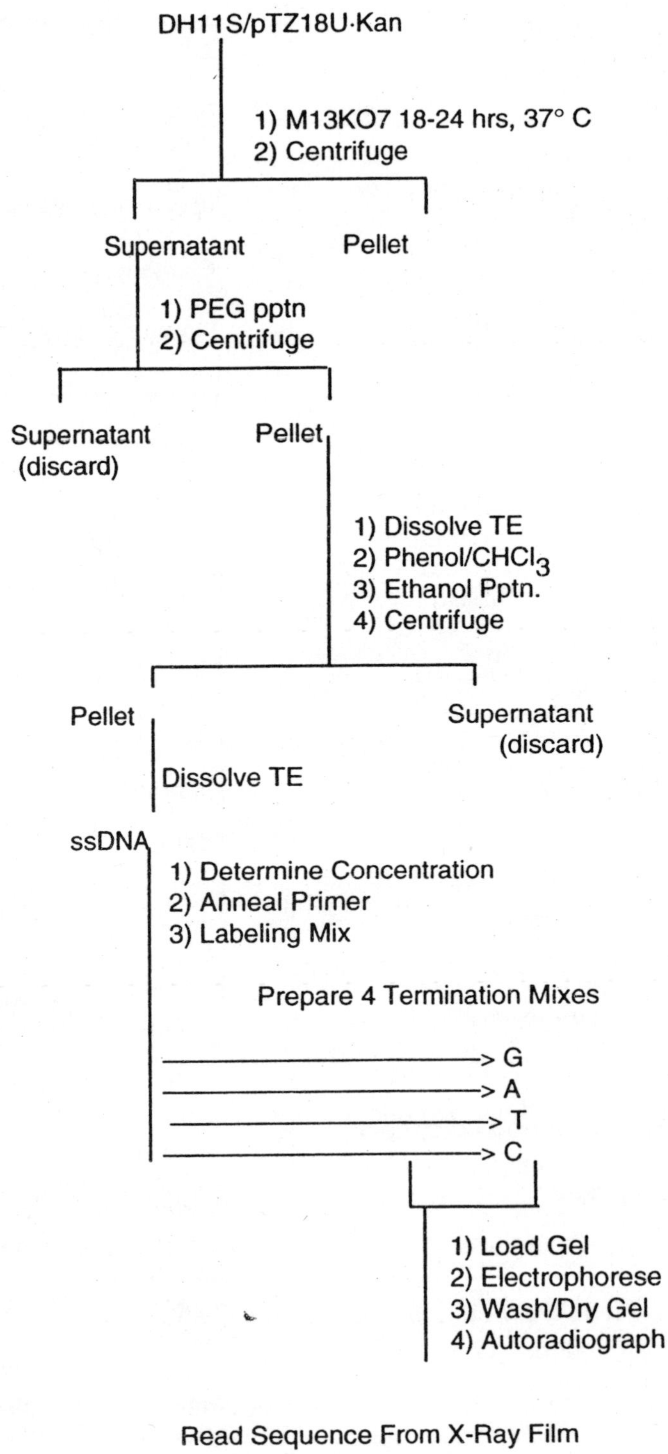

Safety Guidelines

Use gloves and goggles when handling the phenol/chloroform mixture. Phenol causes burns when in contact with skin. Avoid breathing or contact with chloroform. Acrylamide is a neurotoxin. Use a chemical fume hood when weighing dry samples. Wear gloves. Wear gloves when handling acrylamide solutions. Wear gloves when handling polyacrylamide gels. S^{35} is a weak beta emitter. Use the normal precautions for handling radioactive materials. Dispose according to guidelines under the direction of assigned personnel.

Materials

Preparation of Single Strand DNA by Phagemid Method for Sequencing

This exercise will utilize the pTZ18U·Kan recombinant plasmid prepared in Unit 2: Module 8.

DH11S transformed with pTZ18U·Kan cloned within the MCS
TBG medium
  1.2% Tryptone
  2.5% Yeast Extract
  0.4% Glycerol
  12 m<u>M</u> KH_2PO_4
  55 m<u>M</u> K_2HPO_4
  20 m<u>M</u> Glucose
Ampicillin Stock (50 mg/ml)
M13KO7 - (helper phage)
Microfuge Tubes
Pipet Tips
2.5 <u>M</u> NaCl in 40% PEG
TE Buffer
ss-Phenol/$CHCl_3$ (salt-saturated)
3 <u>M</u> Sodium Acetate
95% Ethanol
Gloves
Goggles

Incubator
Shaker Bath
Microfuge
Pipetors

Materials, continued

The Sequencing Reactions

The procedure used here utilizes the Sequenase Version 2.0 kit available from United States Biochemical. Similar kits are available from other companies.

Sequenase Version 2.0
  Sequencing Kit
[a-S^{35}]dATP
Single Strand DNA
Microfuge Tubes
Pipet Tips
Gloves
65°C Water Bath
Pipetors
37°C Bath

Preparation of 8% Polyacrylamide Sequencing Gel

The procedure described here utilizes the BIORAD Sequi-Gen Nucleic Acid Sequencing System. The relatively small size of the gel (21 cm x 40 cm) is convenient for students who have not yet developed the skill to handle a large sequencing gel.

Gloves
Urea
Acrylamide
Bis-Acrylamide
SigmaCote
TEMED
10% Ammonium Persulfate (AP)
10X TBE Buffer
Filter Paper
Pipet Tips (Flat Tip)
Tray 25 x 50 x 10 cm
10% Methanol/10% Acetic Acid
Whatman #1 Filter Paper Precut 21 cm x 40 cm
Plastic Wrap
AGFA-GEVAERT CURIX RP-1 50 AFW 8 x 10" film
Polaroid 6B Filter
Developer
Fixer
Magnetic Stirrer
Electrophoresis Apparatus
2000 V Power Supply
Gel Dryer
Vacuum System
X-Ray Film Cassette

Method

Preparation of Single-Strand DNA by Phagemid Method for Sequencing

1. Plate a dilution (to obtain single colonies) of DH11S transformed with a phagemid vector containing the DNA insert cloned within the MCS. Incubate at 37°C until sufficient growth of colonies has occurred.

2. Suspend a single colony in 2 ml TBG containing 50 µg/ml of ampicillin.

3. Add 10 µl M13KO7 Helper Phage and incubate at 37°C, 275 rpm on a shaker for 18-24 hours.

4. Transfer 1.5 ml of cells to a sterile microfuge tube and spin in a microfuge at maximum speed for 10 minutes.

5. Respin the supernatant in a fresh microfuge tube.

6. Transfer 1.2 ml the supernatant to a fresh microfuge tube and add 300 µl 2.5 M NaCl in 40% PEG 4000. Vortex and incubate on ice for 10 minutes.

7. Centrifuge in the microfuge at high speed for 15 minutes.

8. Remove the supernatant and suspend the pellet in 50 µl of TE buffer. Remove proteins by extracting twice with ss-phenol/ $CHCl_3$.

9. Add 5 µl 3 M sodium acetate and 100 µl ethanol. Place at -20°C for 20 minutes.

10. Centrifuge at high speed for 5 minutes, pour off the supernatant and dry the pellet.

11. Dissolve in 50 µl TE buffer.

Running the Sequencing Reactions

1. Determine concentration of DNA in solution containing DNA to be sequenced. For ssDNA use relationship: 1 A_{260} = 37 µg/ml. The concentration should be approximately 15 µg/µl.

   Keep tubes capped to prevent evaporation.

2. ANNEALING REACTION. DNA polymerases require primer to catalyze incorporation of bases complementary to template strand. In this step, primer is annealed to the template.

Method, continued

In a 1.5 ml tube, mix 1 µl of primer, 2 µl of reaction buffer and 7 µl of DNA solution. Cap the tube, place it in a 65°C water bath and allow the bath to cool to room temperature. The water volume should be chosen such that the cooling requires approximately 30 minutes.

While waiting for the annealing reaction, proceed with steps 4, 5, and 7.

3. After the annealing reaction is complete, place the tube on ice until needed.

4. In a second tube labeled "Diluted Labeling Mix", dilute 4 µl of Labeling Mix with 16 µl of ddH_2O.

5. In a third tube labeled "Enzyme", dilute 1 µl Sequenase enzyme with 7 µl ice cold Enzyme Dilution buffer. Keep on ice.

6. LABELING REACTION. Chain extension from a primer takes place to incorporate radioactively labeled ATP. A broad distribution of (labeled) chain lengths is produced.

   To the tube (from step 2) containing the template and annealed primer, add 1 µl DTT, 2 µl "Diluted Labeling Mix" (from step 4), 0.5 µl [α-S^{35}]dATP, and 2 µl diluted Sequenase enzyme (from step 5). Mix and incubate at room temperature for 2-5 minutes.

7. Label four tubes as "G", "A", "T", and "C", respectively. Add 2.5 µl of the respective Termination mixes to the tubes, i.e. 2.5 µl ddGTP mix to "G", 2.5 µl ddATP to "A", etc. Incubate at 37 °C for at least 1 minute.

8. TERMINATION REACTION. In this reaction, the labeled fragments from the LABELING REACTION are further extended in the presence of all four dNTP's and one of the ddNTP's until each fragment's growth is terminated by incorporation of the ddNTP. Each tube, containing a different ddNTP, will contain only those fragments terminated by the ddNTP present in the particular tube. To each of the four Termination mixes (step 7), add 3.5 µl of [α-S^{35}]dATP labeled DNA (from step 6).

9. Continue incubation for a total of 3-5 minutes.

10. Add 4 µl of Stop Solution to each Termination reaction (from step 8) and store on ice until needed to load sequencing gel.

Method, continued

Preparation of 8% Polyacrylamide Sequencing Gel

1. Into a 200 ml beaker, mix 48 g urea and 40 ml dd H_2O. Heat and stir at 37°C until urea dissolves.

2. Add 7.6 g acrylamide, 0.4 g Bis-acrylamide and 10 ml 10X TBE buffer.

   10X TBE Buffer: Per liter, mix 1.35 M Tris base (163.5 g), 0.45 M Boric Acid (27.8 g), 50 ml 0.5 M EDTA pH 8.0. Final pH approximately 8.9

   CAUTION: ACRYLAMIDE IS A NEUROTOXIN. ALWAYS WEAR GLOVES AND AVOID BREATHING DUST.

3. Stir with stirring bar until dissolved. Filter through Whatman paper No. 1 into a 100 ml graduated cylinder.

4. Adjust final volume to 100 ml with dd H_2O. Store refrigerated in a dark bottle.

Pouring the Gel

Prior to assembling the gel electrophoresis apparatus, coat the inner surface of one of the glass plates with a substance such as SigmaCote (Sigma Chemical Co.) to facilitate release of the gel from the plate at the completion of the electrophoretic run. Also, be sure that the inside surfaces of the gel plates are fastidiously clean so that the polymer solution will flow evenly within the space during the pouring operation.

5. Assemble the gel apparatus. If a sharktooth comb is used, place the comb upside down between the plates so that a level surface is formed at the top of the gel during the polymerization of the gel. Just before loading the gel, the comb can be reversed to properly form the wells.

   When inserting the comb for the first time, be sure to determine how far to insert the comb. If the comb is placed too deep into the sandwich, it will not be possible to load the lanes. A depth of 5-7 mm is adequate.

6. Prepare a plug gel by combining 10 ml of acrylamide/bis/urea solution, 50 µl of TEMED, 50 µl of 10% ammonium persulfate (AP). Mix and pour immediately onto the paper strip in the casting tray.

Method, continued

7. Place the assembled gel plate apparatus in the casting tray and allow the gel solution to rise into the space between the plates at the bottom of the apparatus by capillary action. This should form a uniform gel plug in the bottom of the sandwich. Allow to stand approximately 10 minutes to complete polymerization.

   If using some other type of gel apparatus, a common alternative to this procedure is to securely tape the edges of the sandwich so that no leakage will occur during the pouring of the gel. The tape is removed just prior to setting up the apparatus for electrophoresis.

8. To 50 ml of acrylamide/bis/urea solution, add 0.25 ml of 10% AP and 50 µl of TEMED. Mix and proceed immediately to inject the mixture evenly into the space between plates using a large syringe or squirt bottle. During this procedure, keep the plates at about a 45/10 compound degree angle and avoid trapping any air bubbles. Continue pouring until the space between the plates is full of gel mixture and the mixture is in complete contact with the comb.

9. Allow to polymerize (1-3 hours). Do not attempt to prepare the gel significantly in advance of the time of electrophoresis since only slight deterioration of the gel will lead to unreadable sequence ladders.

Electrophoresis

10. Set up the apparatus in electrophoresis mode. Fill both buffer chambers with 1 X TBE buffer. If a rectangular comb was used in forming the wells, remove it carefully. If a sharktooth comb was used, remove it from the sandwich and reinsert the comb with teeth in contact with top of the gel. Pre-electrophorese for about 30-60 minutes. This can conveniently be done during the time the sequencing reactions are carried out. The gel should reach a temperature of about 45-55°C. This usually requires 1500-1800 V.

11. After the sequencing samples have been prepared (see above), heat the samples to 75-80°C for 2 minutes to denature. Before loading, rinse each well by gently squirting buffer from a tuberculin syringe to which a flattened tip has been attached. The rinsing removes accumulated acrylamide and urea from the solution immediately above the well.

12. Load 2.5 µl of "A" into one lane, then 2.5 µl of "G" into an adjacent lane, then "C", then "T". Special flattened tips for this purpose can be purchased.

Method, continued

13. After the lanes are loaded, set the voltage at approximately 1000 V until the tracking dyes have entered the gel. Gradually raise the voltage to approximately 1500 V or until the gel is running at 45-55 °C. (Sequencing gels are run warm and in the presence of urea to prevent annealing of the single-stranded DNA fragments.) Allow gel to run until the lower tracking dye (bromophenol blue) has migrated 8-10 inches (approximately 45 minutes).

14. Disconnect all electrical connections.

Washing the Gel

When using S^{35} for labeling, the gel must be rinsed and dried. Otherwise, quenching of the weak signal will prevent detection of bands during the autoradiography step.

15. Applying an even force between the plates, carefully separate the plates of the apparatus, being watchful that the gel remains attached to one plate for support.

    THE GEL IS VERY FRAGILE.

16. Place the gel/plate above a bath of approximately 2 liters of 10% acetic acid, 10% methanol. Plates placed directly into the bath have a tendency to separate from the gel which increases the difficulty of subsequent handling. Bathe gel by gently flooding surface for about fifteen minutes (for a 0.4 mm gel).

    BE CAREFUL NOT TO ALLOW THE GEL TO SEPARATE FROM THE PLATE.

17. Carefully remove the gel/plate from the bath and place on the lab bench. Lower a piece of pre-cut (slightly larger than the gel) Whatman No. 1 paper over the gel, being careful not to let air bubbles become trapped between the gel and the paper. If so, gently roll them out with a glass rod or other such device. Slowly and gently lift the paper at one end. The gel should stick to the paper and release from the glass plate. If one end does not release, try the other. If neither end releases readily, carefully loosen the end of the gel with a spatula and try the pickup again.

18. After the gel has been removed from the plate, cover it with a layer of plastic wrap. Trim excess paper and plastic wrap with a single-edge razor blade and straight edge.

Method, continued

Drying the Gel

19. Place the gel on the gel dryer platform with the filter paper on bottom and plastic wrap on top. Close the lid and dry at 80 °C under vacuum for 10-15 minutes or until the temperature is a uniform 80°C across the surface.

Exposing the Film

Depending on the size of the X-ray film used for autoradiography, it may be necessary to further trim the dried gel to size with scissors. AGFA-GEVAERT CURIX RP1 50 AFW 8 x 10 inch film is a convenient size to fit the gel.

20. Using a photographic dark room and a safe light (polaroid 6B filter), load the gel and x-ray film into an x-ray film cassette. Allow 72 hour exposure at room temperature.

Developing the Film

21. In the darkroom, using the safe light with the 6B filter, remove the film from the cassette and soak in developer for 3-5 minutes or until bands become sharp. Rinse in fresh water 3-5 minutes. Soak in fixer 3-5 minutes.

Results

The DNA sequence is read off the film.

UNIT 5: ADVANCED PROTOCOLS FOR DNA ANALYSIS

MODULE 18: PCR AMPLIFICATION

Peter B. Woodruff

* Introduction
* Safety Guidelines
* Materials
* Method
* Results

Introduction

The polymerase chain reaction (PCR), and its alternatives, represent a classic case of technological innovation driving scientific advancement. PCR's impact derives from its remarkable capacity to target a specific DNA sequence amidst great excesses of competing DNA and selectively amplify that sequence perhaps a million-fold or more. The selectivity derives from the judicious choice of primers. The tremendous amplification results from the cyclic repetition of the same DNA polymerase-catalyzed reaction.

If one were to consider whole genomic DNA as a haystack, PCR finds not just a hidden needle but a particular straw. It then produces bales and bales of straw identical to the first. The "straw" need not be particularly free of other debris. Conditions which suit activity of the DNA polymerase component usually support PCR. Source material for amplification range from dried blood or semen through individual human hair or thawed woolly mammoth tissue to mummified brain tissue or fossil remains.

New applications for this powerful, elegant technique seem to appear every week including: 1) genetic mapping, 2) genetic polymorphism, 3) detection of mutations, 4) molecular virology, 5) transcriptional splicing and regulation, 6) microbiology and human disease, 7) molecular archeology, 8) forensic science, 9) population biology and 10) phylogenetics.

Exciting as academic results may be, potential payoffs from diagnostic applications has driven the recent explosion in the development of other methods for cyclic amplification of nucleic acids.

PCR in General

Most applications of PCR rely on the judicious choice of a pair of oligonucleotide primers which are more or less complementary to short stretches (usually 20-24 nucleotides) of DNA bracketing the region of interest and on opposite strands (see "Cycle 1" of Figure 1). In practice, the starting material may be single-stranded DNA (eg: cDNA, c=complementary), melted double-stranded DNA, or even RNA. Given its absolute requirement for a double-stranded starting point, DNA polymerase ideally binds to just one site on each DNA strand. Starting with the 3' end of each primer, the DNA polymerase extends the double-stranded DNA across the region of interest and beyond the complement of the other primer's attachment site. Melting of these double- stranded products yields both "old" and "new" DNA template, each with the appropriate primer binding sites flanking the region of interest. Cyclic repetition of the melting, primer annealing and strand extension steps in the presence of active DNA polymerase, excess primers and nucleotides results in an geometric increase in the DNA sequence framed by the two primers.

Introduction, continued

Schematic of Polymerase Chain Reaction (PCR)

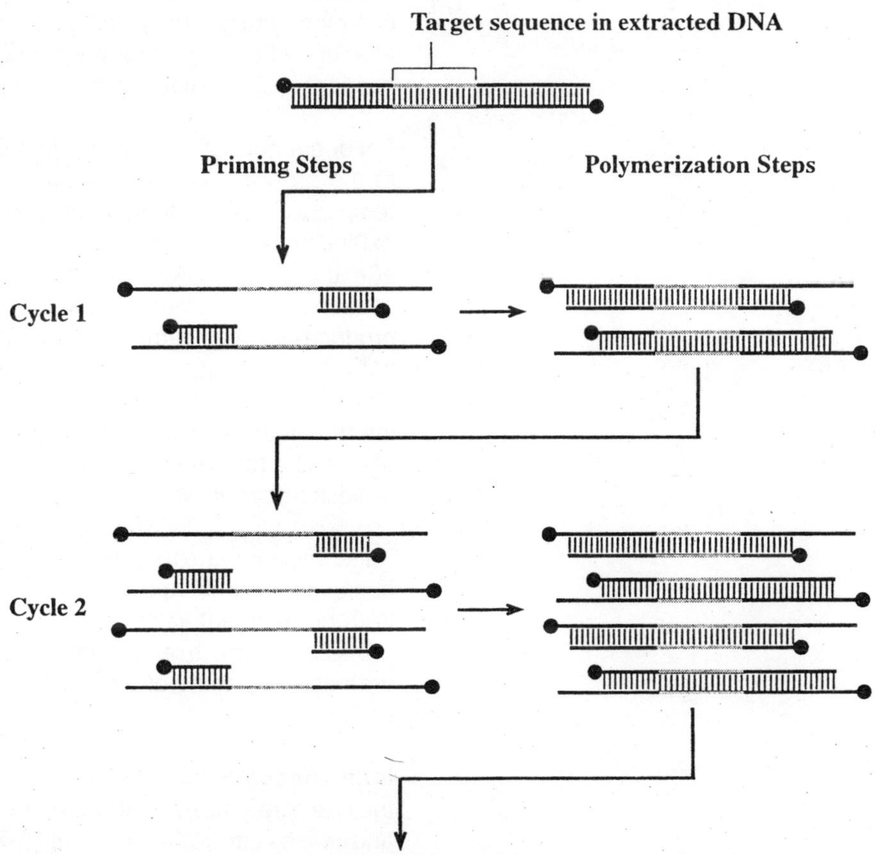

Cycle 3: 8 copies
Cycle 4: 16 copies
Cycle 5: 32 copies
Cycle 6: 64 copies
Ad infinitum

Figure 1

The PCR process is covered by patents owned by Hoffman-LaRoche, Inc.

Introduction, continued

Screening versus Amplification

The first few of these heating and cooling cycles differ subtly from the rest in terms of the templates present. During the first cycle of this screening phase the primers in effect search the starting material for their complementary sequences. Even imperfect complementarity may result from local annealing, polymerase attachment and subsequent strand extension. Note that this may occur at many isolated sites throughout the starting material.

For sequences flanked appropriately by a pair of primer sites, half of the products of the first cycle are the so-called "intermediate templates," each defined at one end by a primer and usually extending beyond the complement of the second primer site. Only after the third cycle do "short templates" of PCR product, demarked at each end by primer sites, begin to appear. Short templates each produce two short templates for each PCR cycle, growing exponentially during the subsequent amplification phase to form the detectable PCR product. In contrast, semiconservative replication of the intermediate templates yields both intermediate and short templates; the former increase linearly and are soon swamped by the rapid increase in short templates.

False intermediate templates, produced in the first cycle at extraneous sites, also disappear in the profusion of true short template. With no associated second site, no short template equivalent is produced. The true selective power of this method lies in this dramatic outpacing of potential competing sequences by the PCR product.

In the late cycles the accumulation of short product eventually slows the rate of reaction. At about 10-8 M of product the reaction is linear and ceases altogether at about 10-7 M. Primers and dNTPs remain in considerable excess. Typically the molarity of the product passes that of the polymerase after perhaps twenty to twenty-five cycles. Eventually self-annealing of the strands or lack of available enzyme becomes limiting. Further cycling beyond the point of saturation leads to the amplification of spurious targets.

While PCR often performs flawlessly, under certain conditions it will amplify undesirable products which may be difficult to distinquish from the intended molecule. If the starting material is heterogeneous within the site being amplified (eg: two or more alleles of a gene are present) or if errors occur during the early rounds of DNA synthesis, multiple products will be produced which may band together on subsequent electrophoresis.

Given the minute amounts of starting material, contamination is a constant concern. Proper technique, including use of frequent

Introduction, continued

glove changes, dedicated positive-displacement pipetors or special "filter tips," prealiquoted reagents and the relative isolation of the PCR preparation and product analysis areas, minimizes the chance that the DNA in small flakes of skin or residue from previous PCR runs could accidently be present in the reaction mixture, yielding spurious results. Several additional approaches have been developed to prevent product carryover from previous PCR reactions.

Temperature

The temperature optima of the melting, annealing and extension steps differ substantially, so temperature control plays an important part in the success of PCR. With the widespread use of thermostable DNA polymerases from thermophilic bacteria, notably Taq polymerase I from the hot spring bacterium Thermus aquaticus, one can practically disregard enzyme lability in the choice of temperature for any step. Two consequences have been especially important: 1) increased control over the stringency of reaction conditions (raising the annealing and extension temperatures reduces nonspecific binding of primers and DNA, thereby reducing background synthesis) and 2) automation of an otherwise labor-intensive procedure.

Melting is usually done at 94°C and annealing at around 45-52°C. The precise choice of annealing temperature controls the stringency of the reaction: 50-52°C is chosen if the primers are known to closely match their complementary sites; 48°C is appropriate if there are perhaps 4-5 mismatches and 45°C produces low stringency conditions for "fishing expeditions".

Protocols differ more widely on the setting for the extension step. The first reports used the heat-sensitive Klenow fragment of *E. coli* polymerase I for the extension step, necessitating the addition of new enzyme after each melting step. Most current PCR protocols specify thermostable DNA polymerases.

Lower temperatures may be specified in some procedures to ensure that the primers remain attached to template strands; if the matching of primers and template is not especially good, denaturation begins before the temperature reaches 72°C , interfering with strand extension. Indeed, annealing and extension may be and often are performed at the same temperature without penalty.

In most PCR laboratories, sophisticated automated thermal cyclers eliminate the otherwise arduous task of switching from one temperature setting to the next over twenty to forty or more cycles. Usually the thin-walled reaction vessel, containing 20-200 µl of reagents overlaid with mineral oil or wax to prevent refluxing, sits

Introduction, continued

with its mates in a form-fitting block which is heated and cooled following preprogrammed instructions. In some thermal cyclers refluxing is minimized by heating the top of the tube, sidestepping the requirement for mineral oil.

Reagents

Let us consider the components of PCR in more detail: 1) the desired template, 2) the primers, 3) the DNA polymerase, 4) the four deoxyribonucleotide phosphates dNTPs and 5) the reaction buffer.

Template

PCR is usually used to amplify DNA template sequences in the 100 to 1000 nucleotide range, though longer stretches are attempted. Reverse transcriptase may be used to generate cDNA (complementary DNA) starter material from a mRNA source. To improve amplification efficiency, circular DNA is usually linearized using a single cutting restriction endonuclease.

Primers

Synthetic oligonucleotides, optimally 20-24 bases in length, must be chosen such that, when bound to opposite strands of the target DNA, their 3'-OHs face each other across the intervening stretch which includes the sequence of interest. The narrow size range of primers reflects a balance between the ease of accurate oligonucleotide synthesis, the requirement to have a sequence long enough to be virtually unique and one long enough to form a stable double-stranded structure to act as a primer.

Primer-target DNA base pairing must be especially precise at the 5 to 6 bases at the 3'-OH end to promote effective amplification, although some variations on the technique make use of the much laxer 5' end requirements to add in base sequences not found in the template, such as restriction endonuclease recognition sites, or other molecules, such as biotin labels for detection purposes.

Primers having self-complementary regions which form secondary structures such as hairpin loops should be avoided, as should primer pairs showing mutual complementarity at or near their 3' ends. The latter may anneal, forming a relatively short polymerase template. Once extended to its double-stranded form by DNA polymerase, it becomes an undesirable primer-dimer which may out-compete the desired product.

Introduction, continued

High primer concentration may cause mispriming and the synthesis of undesired nontarget sequences; low primer concentration will limit the degree of amplification possible. Primers with even G, A, T and C composition and without repeats are desirable.

Alternative Method for Amplifying Nucleic Acids

Nucleic acid amplification methods generally fall into one of two classes:

1) those that select particular stretches of DNA or RNA and enzymatically copy them in vast numbers (PCR, nucleic acid based amplification (NASBA™), transcription based amplification (TAS) etc.);

2) those that detect particular stretches of DNA or RNA and enzymatically amplify a form of the probe used in the detection process (Qb replicase amplification, ligase chain reaction [LCR], cycling probe reaction etc.).

Both NASBA™and TAS rely on the ability of RNA polymerases to make multiple copies (10 to 103) of selected templates per cycle, reducing the number of cycles necessary for million-fold amplifications. Both can also start with either DNA or RNA sequences. Cangene's NASBA™ operates at 42°C without thermal cycling.

Qb replicase amplification also relies on the multifold amplification of RNA, this time by the viral RNA-dependent RNA polymerase Qb replicase (Qb replicase is a tetramer. One subunit is coded for by the small Qb RNA; three of the subunits are actually bacterial polypeptides recruited for the purpose. A class of recombinant RNA probes combining a hybridization sequence specific for a target RNA or DNA with the MDV-1 RNA sequence are utilized. The high degree of specificity of Qb replicase for the latter sequence plus the fact that both the template and product nucleic acids are free to act as templates in the next round of synthesis results in exponential increases in probe molecules. Key to the success of this technique, as for other probe-based approaches, is effective removal of unbound probe molecules, which would give false signals. Although the reaction cycles in each round of product produces template for yet another round, there is no requirement for thermal cycling.

The ligase chain reaction depends on the simultaneous annealing of two probes designed to fit side-by-side on the DNA to be targeted. DNA ligase covalently links the adjacent probes. Heating the reaction mixture renders it single-stranded; the joined probes form

Introduction, continued

additional targets for subsequent annealing and ligation. The process can be repeated many times and leads to an exponential increase in the number of target equivalents.

The probe in the cycling probe™ reaction is a DNA-RNA-DNA chimera. At a defined operating temperature appropriate to each probe, the probe will bind to a target DNA sequence. Treatment of the bound probe with ribonuclease H removes the RNA portion, leaving two separate DNA probe fragments. At the chosen operating temperature these shorter DNA fragments fall away from the target DNA. This re-exposes the probe attachment site for another round of attachment, digestion and release. Accumulation of the probe fragments provides the signal indicating the presence of the site. In this process there is no requirement for thermal cycling.

The objective of this module is to demonstrate the amplification of target DNA as a function of the number of amplification cycles. The principles discussed in the previous section of this module are to be demonstrated through the use of a typical PCR recipe. A kanamycin gene insert within a pTZ18U vector will be amplified using M13 universal and Reverse primers to target the kanamycin DNA insert. (See Unit 2, Module 8 for construction of the pTZ18U/Kan vector.)

The procedure assumes the availability of a thermocycler. If such is not the case, three water baths at the given temperatures can be set up. Manual transfers can be made from bath to bath according to the time intervals required. Several tubes can be placed on a suitable holder to facilitate transfer from bath to bath. A simple homemade device has also been reported in the literature.[1]

[1]Watson, Robert. "PCR in a Teacup: A Simple and Inexpensive Method for Thermocycling PCR's" in Innis, M.S., D.H. Gelfand, J.J. Sninsky and T. White (eds.) PCR Protocols: A Guide to Methods and Applications, Academic Press, 1990.

Safety Guidelines

Boiling agarose can spatter and cause severe burns. When heating agarose wear safety goggles and hot gloves. Latex gloves should be worn throughout the procedure, especially when handling solutions or gels containing methylene blue. Prior to turning electrophoresis power pack on, be sure that the chamber is level and that the work surface is dry. Wear gloves and safety goggles when operating electrophoresis apparatus.

Ethidium bromide is a powerful mutagen. Wear eye and skin protection (goggles/gloves) while working with ethidium bromide to protect from UV radiation emitted from the transilluminator. Ethidium bromide solution should be treated as hazardous waste and disposed of properly.

Materials

Amplification Buffer (10X)
- 0.5 M KCl (37.5 mg/ml)
- 0.1 M Tris· HCl pH 8.4 (15.8 mg/ml)
- gelatin (1 mg/ml)
- $MgCl_2$ (2.5 mM)
- sterile ddH_2O to volume

2 mM 4 dNTP Mix
M13 Universal Primer
M13 Reverse Primer
Template DNA (pTZ18U/Kan)
100 mM $MgCl_2$
Taq DNA Polymerase
Thermocycler Tubes
Agarose
TBE Buffer
Ethidium Bromide (1 mg/ml)
Polaroid Film 667 (optional)

Thermocycler or 3 Water Baths
Pipets
Electrophoresis Cell
Power Supply
Transilluminator
Camera (optional)

Method

The $MgCl_2$ concentration (2.5 mM) is suitable for the following procedure. In other systems it is advisable to determine the optimal concentration in the amplification buffer by employing several simultaneous trial runs varying the $MgCl_2$ concentration from 1.5 to 10 mM. Stock solutions of sterile 1.0 M $MgCl_2$ can be purchased from Sigma Chemical Co.

1. Prepare the following mixture for each 0.5 ml PCR tube:

   10 µl 10x Amplification Buffer (+ optimized $MgCl_2$)
   10 µl 2 mM 4 dNTP Mix
   10 µl 5 µM M13 Universal Primer
   10 µl 5 µM M13 Reverse Primer
   1 ng Template DNA
   ddH_2O to 99 µl
   1 µl 2.5 U/µl Taq DNA Polymerase

2. Remove 5 µl and save for gel analysis.

3. Cover the mixture in the 0.5 ml tube with 100 µl mineral oil to prevent evaporation. Alternatively, specially designed PCR tubes are available through Barnstead/Thermolyne Corp. These tubes hold 100 µl with no head space and do not require adding mineral oil. The disadvantage of mineral oil is that it makes it difficult to cleanly remove aliquots.

4. Set up a thermocycler program as follows:

   94°C - 90 seconds
   55°C - 120 seconds
   72°C - 60 seconds
   Repeat 4 cycles and remove 5 µl aliquots.

5. Continue 4-cycle repeat, sample removal until 20 cycles have been completed.

6. Prepare a 0.8% agarose gel for analysis of the 5 µl aliquots. Dot the aliquots onto the gel and allow the DNA to penetrate the gel over a period of one hour. Stain the gel with ethidium bromide and view using a transilluminator. An alternative is to stain the gel with ethidium bromide prior to depositing the 5 µl aliquots. Observe the increased intensity of the dots as a function of the number of cycles. Optionally, photograph the gel on the transilluminator using Polaroid 667 film.

Results

The amplified DNA will be visualized as a more intensely stained band over increased cycles as demonstrated on the agarose gel.

UNIT 5: ADVANCED PROTOCOLS FOR DNA ANALYSIS

MODULE 19: PREPARATION OF BIOTINYLATED PROBES

Anthony Sena
Harley Mortensen

* Introduction
* Safety Guidelines
* Materials
* Method

Introduction

The ability to analyze DNA fragments in a specific and repeatable manner was a keystone event in the advancement of molecular biology and the development of biotechnology. This was made possible by the discovery of bacterial enzymes that "cut" double stranded DNA at specific locations within the molecule. The natural function of such enzymes is to protect bacterial cells from attack by viral DNA. Classified as restriction endonucleases (or restriction enzymes) these enzymes recognize, bind to, and cut at specific nucleotide sequences known as recognition sites. As is the case with many other proteins that interact with DNA, these recognition sites generally display twofold symmetry. For example, *Eco* RI, isolated from the bacterium *Escherichia coli*, cuts the DNA phosphodiester bond between the guanine and adenine base in the nucleotide sequence 5'-GAATTC-3'. In double stranded DNA the base sequence GAATTC will base pair with the same sequence in an antiparallel direction. *Eco* RI therefore cuts both DNA strands between the G and the A bases. There are hundreds of restriction enzymes, most recognizing different cutting sites. Restriction endonucleases are classified with three letters derived from the scientific name of the bacterium that produces the enzyme. For example, *Eco* RI from *E. coli*, *Sac* I from *Streptomyces achromogenes*, and *Sal* I from *Streptomyces albus*.

As is the case with most genetic characteristics, restriction sites are polymorphic between individuals. This means that different individuals have differing restriction enzyme recognition sites within their DNA. When DNA from different individuals is digested by restriction enzymes the result is differing sets of DNA strands known as Restriction Fragment Length Polymorphisms (RFLPs). RFLPs occur because of mutations and polymorphisms in DNA. In some cases mutations are deleterious and are manifested in disease, dysfunction, or death. Other mutations result in harmless, but differing, characteristics among individuals. The result of such mutations and polymorphisms is populations comprised of individuals with highly variable and unique nucleotide sequences. RFLP analysis allows us to study this variability. Suppose that a mutation in a restriction site occurs within a DNA sequence that codes for part of a protein. Because of the nature of the genetic code, a mutation need not result in a different protein being expressed. Figure 1 below shows how such a "silent" mutation can eliminate an *Eco* RI recognition site (GAATTC) and not change the expression of the genetic code.

Introduction, continued

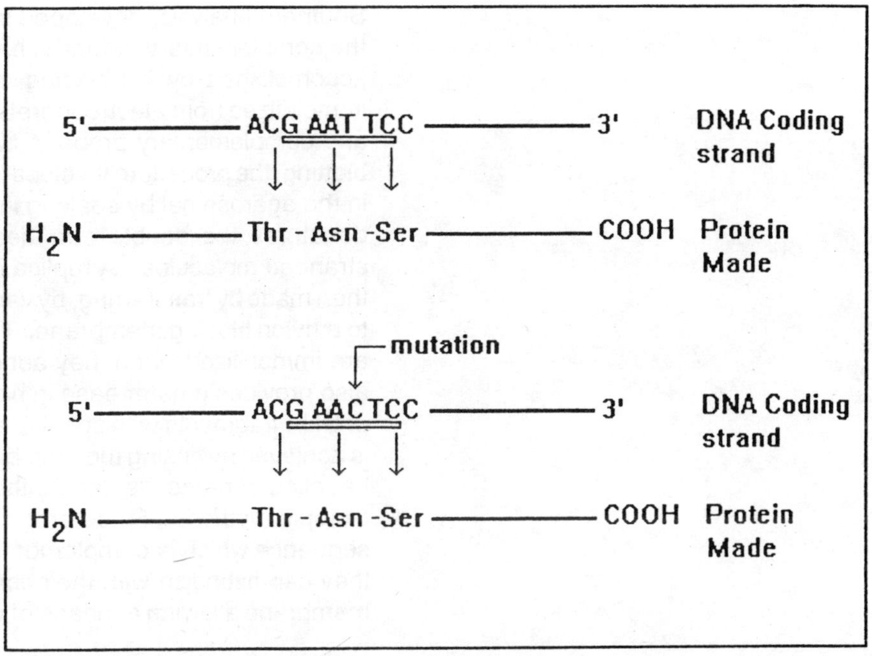

Figure 1: Mutation of *Eco* RI site without phenotypic change

DNA base sequences which do not code for any protein make up a significant percentage of the eukaryotic genome. The absence or presence of restriction enzyme recognition sites in such sequences leads to a high degree of RFLP variability among individuals. These variations can be easily detected through RFLP analysis. More importantly, RFLPs are stable and heritable characteristics which can be studied in the same manner as classical alleles. For this reason RFLP analysis has become the basic, diagnostic test for differences between individuals. Also, RFLPs are important in basic research as markers in linkage analysis. In addition to its utility in identity testing ("DNA fingerprinting") in paternity and criminal cases, DNA analysis is now used for a wide range of applications from screening for genetic diseases to the localization and characterization of many human genes. The genome of most eukaryotes is comprised of a large number of base pairs (3 billion bp in humans). This large size allows for the presence of many recognition sites for any particular restriction enzyme. For example, restriction enzyme digest and gel electrophoresis of eukaryotic genomic DNA will yield so many differently sized fragments that the cleaved DNA will appear as a continuous smear on an electrophoretic gel with few, if any, discrete banding patterns.

Introduction, continued

Southern analysis, developed in 1975 by E. M. Southern, exploits the complementary nature of base pairing in nucleic acids. This is accomplished by hybridizing denatured RFLPs that have been immobilized from electrophoretic gels with labeled, single stranded and complementary probes. Sometimes referred to as Southern blotting, the procedure involves the denaturation of RFLP fragments in the agarose gel by soaking in NaOH (alkali). The alkali solution denatures the double stranded DNA fragments into the single stranded molecules. A replica of the pattern of DNA fragments is then made by transferring, by various methods, these single strands to a nylon blotting membrane. The single stranded DNA fragments are immobilized when they adhere to the nylon membrane, which also provides greater ease in handling than the gel. The detection of DNA fragments with specific sequences on the nylon membrane is achieved by rinsing the membrane in a solution of labeled probes. Labeling of probes is done either by isotopic (radioactive) or non-isotopic methods. Since probe molecules also have a nucleotide sequence which is complementary to a particular base sequence, they can hybridize with their complementary strands on the nylon membrane allowing a means of visualizing discrete RFLP patterns.

Using isotopic probes, visualization occurs when photographic film is placed over the membrane and radiation from the probes exposes the film, revealing the location of the target DNA. Because of the hazards associated with radioactive handling and disposal, alternative methods of labeling probes have been developed. Two of the modules in this unit describe the preparation of non-isotopically labeled probes using biotin as the label. One method involves incorporation of the label through a photochemical reaction while the other utilizes the process of nick translation, an enzymatic method.

Introduction, continued

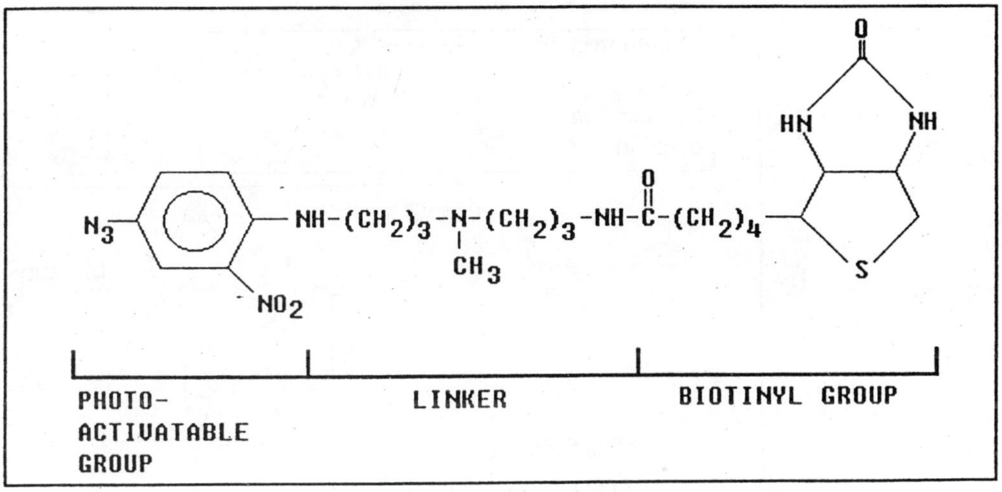

Figure 2: Structure of photobiotin.
Adapted from Focus, 19(4), 1987, BRL/Life Technologies, Inc.

Labeling Probes by Photochemical Activation of PAB

Biotin, one of the fat-soluble vitamins, forms very stable complexes with the protein avidin. Biotinylated probes, when paired to their complementary target strands, are detected by treatment with avidin-alkaline phosphatase conjugates. The DNA-biotin-avidin-alkaline phosphatase complex is made visible by exposure to a mixture of 5-bromo-4-chloro-3-indolyl phosphate (BCIP) and nitro blue tetrazolium (NBT). Alkaline phosphatase catalyzes the hydrolysis of phosphate from BCIP which undergoes a redox reaction with NBT to produce a blue precipitate at the site of the reaction. Photoactivatable biotin (PAB) allows preparation of probes non-enzymatically. PAB (Figure 2) is available as a lyophilized solid which is reconstituted with distilled water to a stock solution concentration of 1 mg/ml.

The reaction involves exposing the probe DNA and PAB to light from a standard heat lamp for a short period of time, extraction of unreacted PAB with 2-butanol, and ethanol precipitation of DNA (Figure 3).

Introduction, continued

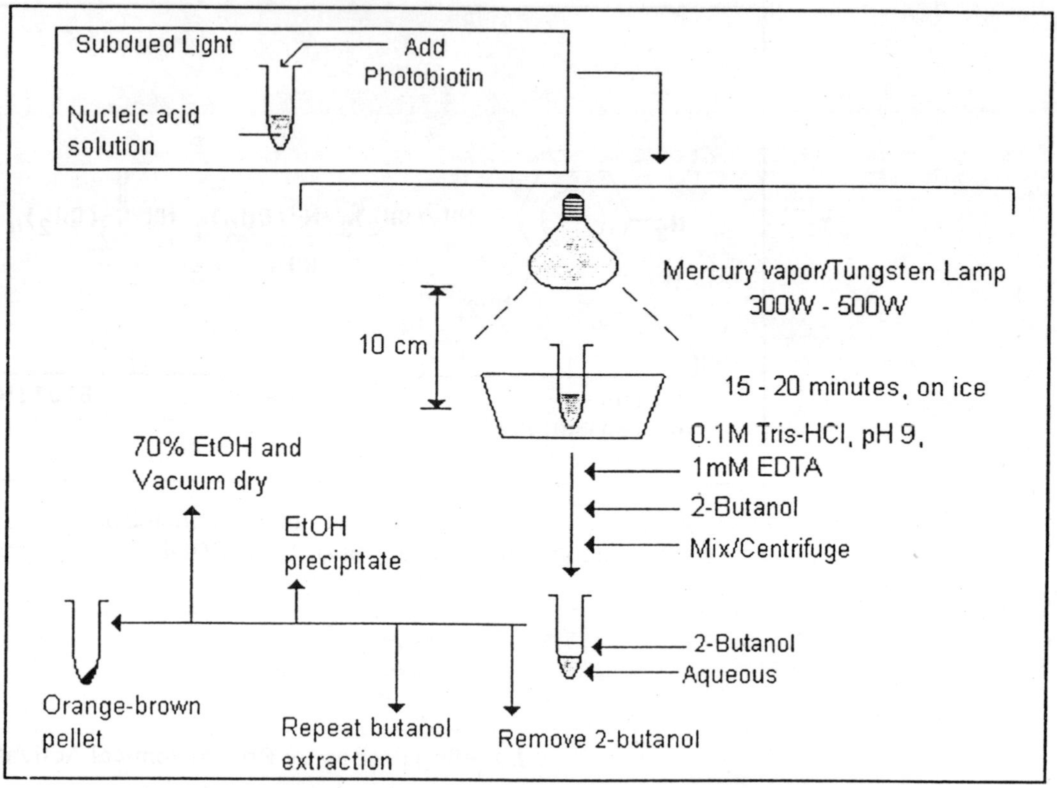

Figure 3: Procedure for photobiotin labeling of nucleic acids

TE buffered probe solutions so prepared can be stored refrigerated in the dark for several months, in contrast to radiolabeled probes.

DNA/PAB reactions are run at mass ratios of 1:1-3. The procedure described below employs a 1:2 DNA:PAB ratio.

Nick Translation Labeling of Probe DNA

The nick-translation reaction can also be used to incorporate biotinylated nucleotides into an oligonucleotide probe. In this procedure, double-stranded oligonucleotides are labeled during incubation in an optimized buffer containing DNase I, DNA polymerase I, and dNTP's (Figure 4). Generally, one of the four dNTP's in the mixture is labeled with biotin. Biotin labeled nucleotides react with strepavidin alkaline phosphatase (SAAP), an enzyme produced by the bacterium *Streptomyces avidini*, to form a covalently linked complex of DNA-biotin-strepavidin. Exposure of the biotin-

Introduction, continued

SAAP complex to a solution of the substrate 5-bromo-4-chloro-indolyl phosphate (BCIP) and the dye Nitro Blue Tetrazolium (NBT) will act to reduce the NBT to an insoluble blue product, revealing the location of target-probe DNA hybrids on the nylon membrane.

During the nick translation reaction DNase I produces random "nicks" along the backbone of double stranded DNA to be labeled, exposing free 3' hydroxyl ends within the nicked strand. At the exposed 3' hydroxyl, DNA polymerase I sequentially adds nucleotides from a pool of dNTP's using the template strand for complementary base pairing. The reaction also takes advantage of the 5' -> 3' exonuclease activity of DNA polymerase I to remove the existing strand as it incorporates new nucleotides.

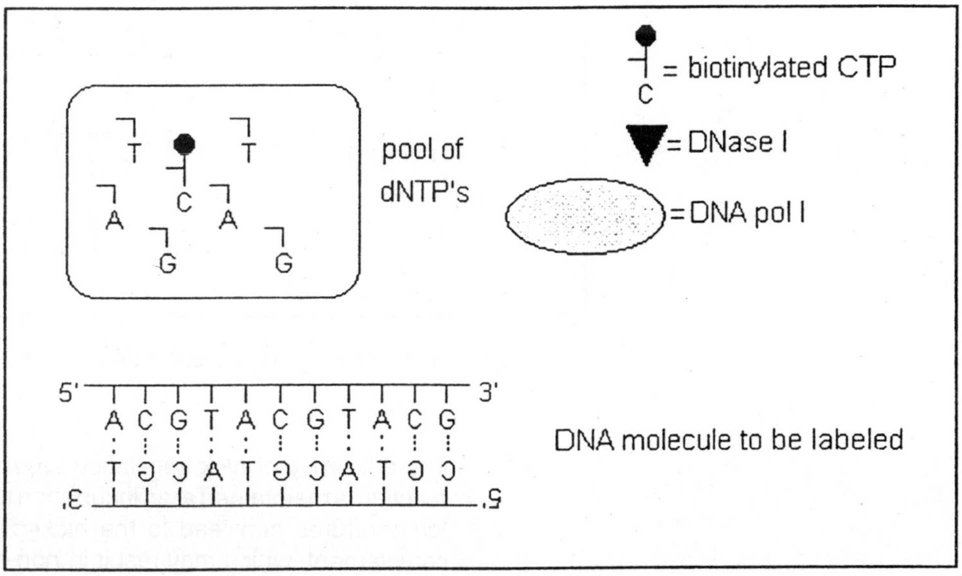

Figure 4: Requirements for the nick translation reaction

Introduction, continued

The presence of the biotinylated dNTP allows for the newly synthesized strand to become labeled (Figure 5).

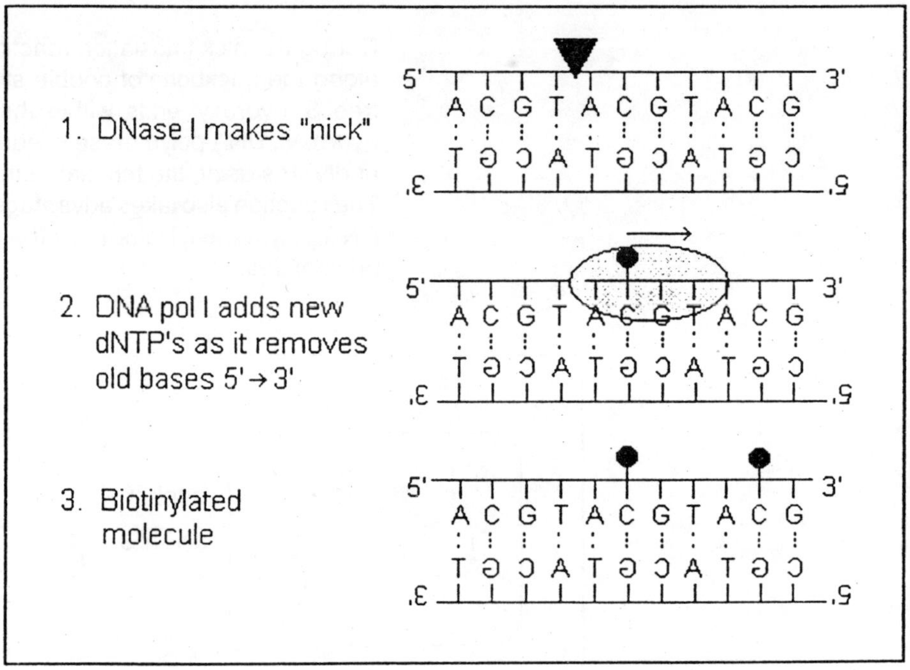

Figure 5: Schematic representation of the nick translation reaction

The efficiency of nick translation labeling is maximized when this reaction is maintained at an incubation temperature of 15°C. Higher temperatures can lead to the nicked strand separating from its complement, which may result in non-specific probe molecules.

Safety Guidelines

Follow routine safety procedures as required by the instructor.

Materials

Labeling Probes by Photochemical Activation of PAB

Lambda DNA (0.5 µg/µl)
PAB stock (1 mg/ml)
ss-2-Butanol (ss-2-Butanol is prepared by shaking 1 volume of 2-butanol in 2 volumes of TE buffer in a separatory funnel. Allow the layers to separate, collect the upper organic layer of ss-2-butanol)
3 M Sodium Acetate
95% Ethanol
TE Buffer
Certain reagents for non-isotopic labeling are available from Clonetech Labs

Microfuge Tubes
Pipets
Tips
Heat Lamp
Microfuge

Reagents for Southern Blotting experiments are available as a kit in Cat. #207 and 311 from EDVOTEK.

Nick Translation Labeling of Probe DNA

Biotin-14-dCTP (0.4 mM)
Carrier RNA (10 µg/µl)
Unlabeled dNTP's (from 1.5 mM stock)
0.2M EDTA Ethanol (70%)
Gloves
Ice
Isopropanol (95%)
Lambda *Hind* III DNA (1 µg/µl)
Nick translation buffer (10X)
500 mM Tris-Cl, 100 mM $MgSO_4$, 1 mM DTT
Optimized DNA polymerase I/DNase I Mix
2M Sodium Acetate (pH 6.5)
TE buffer (pH 8.0)
Sterile ddH_2O

Temperature Controlled Incubator
Microfuge
Pipet
Vortex mixer

Method

Labeling Probes by Photochemical Activation of PAB

1. In a 1.5 ml microfuge tube, place approximately 15 μg of Lambda DNA (usually about 30 μl of approximately 0.5 μg/μl).

2. Add 1.5 μl of 3M sodium acetate and 60 μl 95% ethanol to precipitate the DNA. Chill 20 minutes at -20°C and centrifuge at high speed in a microfuge for 5 minutes.

3. Discard supernatant, dry pellet and dissolve it in 50 μl ddH_2O.

4. Add 30 μg (30 μl) of the reconstituted photobiotin solution. Mix and place in an ice bath with the tube open.

5. Position a heat lamp approximately 10 cm above the tube and shining directly into it. Irradiate in this manner for 15 minutes.

6. Extract with an equal volume of ss-2-butanol to remove unreacted photobiotin. Remove and discard the colored upper organic layer.

7. Repeat step 6.

8. Precipitate the DNA by adding 8 μl 3M NaAc and 160 μl 95% ethanol. Chill at -20°C for 20 minutes and centrifuge to pellet the DNA. Remove the supernatant and dry the pellet by placing the tube inverted over a paper towel.

9. Redissolve the biotinylated Lambda DNA in 50 μl TE buffer.

10. Biotinylated probes prepared in this manner may be stored in the freezer at -20°C for up to one year. The probe prepared according to the directions in this module can be used with a *Hind* III (or other) digest of Lambda to demonstrate Southern transfer (see module 20).

Method, continued

Nick Translation Labeling of Probe DNA

1. Prepare the nucleotide mix by adding equal volumes of unlabeled 1.5 mM stock dNTPs (dATP, dGTP, dTTP) to a final volume of 10 ml.

2. Into a sterile 1.5 ml centrifuge tube add the following to a final volume of 50 µl:

   a. 26.5 µl of sterile triply distilled water
   b. 10 µl of unlabeled dNTP's
   c. 5 µl of 10 X nick translation buffer
   d. 1 µl of l *Hind* III DNA (1 µg/µl)
   e. 2.5 µl of 0.4 mM Biotin-14-dCTP
   f. 5.0 µl of DNA Polymerase I/DNase I mix

3. Incubate at 15°C.

4. Stop the reaction after 1 hour by adding 1 µl of EDTA stop buffer then dilute by adding 150 µl of sterile water.

5. Add 1 ml of carrier RNA + 20 µl of Sodium Acetate + 300 µl of Isopropanol. Vortex.

6. Place in a -20°C freezer for at least 2 hours.

7. Remove the tube from the freezer and return to room temperature for at least 30 minutes.

8. Centrifuge the tube, at maximum speed, for 15 minutes to pellet the nick translated probe.

9. Decant the supernatant and rinse the pellet in 100 µl of 70% EtOH. Centrifuge resuspended pellet for 10 minutes.

10. Decant the supernatant and air dry the pellet. Resuspend in 100 µl of TE.

11. Store at -20°C until needed in the Hybridization/Detection procedure (see module 20).

UNIT 5: ADVANCED PROTOCOLS FOR DNA ANALYSIS

MODULE 20: SOUTHERN BLOT ANALYSIS

Anthony Sena
Harley Mortensen

* Introduction
* Experimental Outline
* Safety Guidelines
* Materials
* Method
* Results

Introduction

Cell host DNA digested using restriction endonucleases often results in a large number of such fragments having a broad size distribution. Electrophoresis of such digests will separate the fragments into a continuous "smear" of little analytical value when visualized. Most applications in biotechnology and forensic analysis of DNA require a specific fragment to be selected from these large continuous collections of fragments.

In 1975, Edward M. Southern introduced a method that makes possible the selection of specific DNA fragments from a large number of electrophoretic fragments. This method involves a capillary transfer of DNA from an electrophoretic gel onto a membrane with a relatively high affinity for single stranded DNA. Under appropriate conditions, both nitrocellulose and nylon membranes have high affinity for single stand DNA, therefore a transfer from a gel can be accomplished by denaturing the double stranded fragments in a manner that maintains the integrity of the fragment position. Exposure of the gel to alkaline solutions of NaOH and NaCl will denature double stranded DNA to its single strands. Capillary transfer occurring in this denaturing solution will "force" single stranded DNA onto the membrane filter. The membrane immobilizes single stranded DNA in the same order as in the gel. Localization of the specific strand is then accomplished by complementary base pairing with a radioactively labeled "probe" of RNA or DNA. The membrane is rinsed to remove any non-hybridized probe and then exposed to a sheet of x-ray film resulting in an autoradiograph. Bands appearing in the autoradiograph correspond with the sites of the specific bands originally in the gel. The autoradiograph serves as a permanent record for analysis and archive purposes.

RNA can be transferred to a membrane in a variation of the Southern blot. This so-called "Northern" blotting allows the detection of small RNA molecules within a cell. Because RNA molecules do not readily bind to nitrocellulose, early blotting transfers were done using a special diazobenzylmethyl (DMB) filter. More recently, special buffers have been developed which allow the transfer of RNAs to nitrocellulose. Northern and Southern blot techniques are essential tools of molecular biology and biotechnology.

Analogous procedures have now been developed to detect specific proteins in SDS-PAGE gels. Unlike Southern or Northern blotting techniques, protein methods do not utilize complementary hybridization. Instead, this procedure known as the Western blot, employs antibodies labeled with radioisotopes or enzyme complexes to probe specific proteins.

The time limiting phase of the Southern, Northern, and Western blots is the transfer of the nucleic acid or protein into the membrane.

Introduction, continued

Recent protocols have been developed that allow the transfers to be made in a vacuum apparatus (vacuum blot) or to be transferred electrophoretically (electroblot). Such time saving protocols, along with the use of non-isotopic probes have made these important tools available to undergraduate and secondary school curriculum.

The procedure below calls for preparation of a lambda *Hind* III digest which will be transferred to a membrane. One method employs a capillary transfer while the other demonstrates an electrophoretic transfer using the Mini Trans-Blot Assembly that is commercially available. The membrane will then be used in for hybridization and detection with the biotinylated lambda probe prepared in Module 19 of this unit.

Experimental Outline

A. Agarose gel electrophoresis of RFLP fragments

Prepare agarose gel — Prepare RFLP DNA

Electrophoresis - 60-90 min

Stain-Destain - 20 min

Visualize, Photograph

↓

B. Set up Southern transfer apparatus — Prepare transfer buffers

Denature gel - 45 min

Remove denat. solution

rinse — Neutralize gel 30 min

Prepare nylon membrane

Set up blotting apparatus

↓

Transfer overnight - Room temp

Safety Guidelines

Ethidium Bromide is a mutagen. Use gloves when handling ethidium bromide. Use goggles or other proper eye protection when viewing through the transilluminator. Ultraviolet radiation from the transilluminator can cause tissue damage. Take appropriate cautions when setting up and using the electrophoresis apparatus and power supply. Potential hazards exist in working with formamide and NaOH. Follow instructors' guidelines.

Materials

Preparation of a *Hind* III Lambda Digest

- Microcentrifuge Tubes
- Lambda DNA (0.5 μg/μl)
- 10X Core Buffer (BRL)
- Sterile ddH_2O
- *Hind* III Lambda DNA
- Tracking Dye
- Agarose
- TBE Buffer
- Ethidium Bromide (10 μg/ml)
- Polaroid Film 667 (optional)
- 37°C Water Bath
- Electrophoresis Chamber
- Power Supply
- Transilluminator
- Camera (optional)

Capillary Transfer of Restriction Fragments

- 10X TBE buffer for electrophoresis
- Stretch vinyl or latex safety gloves
- Denaturation buffer (0.5M NaOH, 1 M NaCl)
- Neutralization buffer (0.5M Tris-HCl, pH = 7.4, 3M NaCl)
- Whatman 3MM filter paper
- Nylon blotting membrane (SIGMA, # N6517)
- Plastic wrap
- Variable (10 ml to 50 ml) pipettes & tips.
- 50 ml and 250 ml beakers
- Sterile Millipore forceps

Electrophoretic Transfer of Restriction Fragments

- TBE Buffer (0.5x)
  (0.4 mM Tris, 0.4 mM boric acid, 1 mM EDTA, pH 8.3) Per liter, mix 5.4 g Tris, 2.75 g boric acid, 2 ml 0.5 M EDTA - (pH 8.0), ddH_2O
- Nylon Transfer Membrane
- .25M NaOH
- Whatman #1 Filter Paper
- Gloves
- Mini Trans Blot Apparatus
- Magnetic Stirrer
- Power Supply
- Vacuum Oven

Materials, continued

Hybridization Materials

20x SSC (3 M NaCl 0.3M sodium citrate, pH 7.0
1% SDS
2x Prehybridization solution (Sigma Chemical Co.)
Sealable Plastic bag
Membrane from Southern Transfer
2x Hybridization Buffer (Sigma Chemical Co.)
Biotinylated probe
42°C Water Batj
Pipets
Pan or Tray
Microwave oven or heat block

Detection Buffers (Uses Clontech Protocol)

Buffer A: 1 M NaCl, 0.1 M Tris-HCl (pH 7.5), 2 MM $MgCl_2$, 0.5% Triton X-100
Buffer B: 3% Bovine serum albumin in Buffer A
Buffer C: 0.1 M NaCl, 0.1 M Tris-HCl (pH 9.5), 20 mM $MgCl_2$

Method

Preparation of a *Hin*d III Lambda Digest

In a 1.5 ml microcentrifuge tube, add the following:

2 µl Lambda DNA (0.5 µg/µl)
5 µl 10X core buffer (React 2)
41 µl sterile ddH_2O
2 µl *Hind* III

Incubate at 37°C for 30-45 minutes. Add 5 µl tracking dye and load 10, 15, 20 µl in separate lanes of agarose gel and electrophorese. Stain gel with ethidium bromide and check results using a transilluminator. Photograph gel if a permanent record is desirable.

Lambda digest of *Hind* III is available from a variety of sources.

Method, continued

Capillary Transfer of Restriction Fragments

1. Soak the gel in 350 ml of Denaturation Solution for 45 minutes. This will separate and denature the double-stranded DNA RFLP fragments into their complementary single strands.

2. Remove the Denaturation Solution, rinse the gel once in distilled water and place it in 350 ml of Neutralization Solution. Allow 30 minutes for neutralization.

3. While gel is in the Neutralization solution, you should prepare the nylon membrane. Always wear gloves when handling nylon membrane, it is fragile and should be handled with care.

4. Wet the nylon membrane in distilled water. Note the two different textures on each side. If necessary, trim with scissors so that it is just slightly greater than the size of the gel. To aid in proper orientation, mark the membrane by cutting a notch at the right hand corner at the edge nearest the wells

5. Place a sheet of plastic wrap on a level bench top. Remove the gel from the Neutralization Solution and place it well side down on the plastic wrap. Invert gel and place it in contact with the smooth surface of the nylon membrane. Make sure the sides of the membrane do not overlap onto the sides of the gel.

6. Place a sheet of filter paper (Whatman, 3MM paper), cut to the same size as the membrane, on top of the membrane. Do not allow the filter paper to hang over the sides of the membrane.

7. Set up blotting apparatus as shown in the figure below. This transfer step should occur overnight at room temperature.

8. After transferring overnight, remove the paper towels, the weight and the filter paper. Make sure that you are wearing gloves and, using a sterile flat forceps, remove the membrane and place it on a dry, clean paper towel with the DNA side up. Use a pencil to label the DNA side of the membrane with your lab and group name. What is left of the gel should be very thin and it can be discarded.

9. Place the membrane between two sheets of Whatman 3MM paper and incubate for 30 minutes at 80°C in a conventional incubator. After incubation, RFLP fragments should be immobilized on the nylon membrane and are ready for hybridization.

Method, continued

Electrophoretic Transfer of Restriction Fragments

1. Cut a nylon membrane to the same size as the gel containing the DNA to be transferred. Use gloves when handling the membrane to avoid fingerprints.

2. Using a soft pencil, devise a code that will allow a determination of the orientation of the filter.

3. Denature the DNA in the gel by soaking the gel in 0.25 M NaOH for 15 minutes. Place the gel in TBE buffer for 10 minutes in order to neutralize the NaOH.

4. Load the transfer apparatus by placing on the gray cassette panel, in sequence, a 0.5X TBE-saturated pad, filter paper (Whatman No. 1 cut to gel size), gel, membrane, filter paper, pad. Be sure that there is intimate contact between the gel and the membrane and that there are no entrapped air bubbles. Do this in a shallow tray so that all components can be well saturated with buffer.

5. Close the cassette latch and place into the buffer chamber (be sure that the orientation of the gel relative to the transfer filter is correct, i.e. the membrane must be on the "anode" side of the gel with gray cassette panel facing gray electrode panel).

6. Insert the cooling unit, place a magnetic stirring bar in the buffer chamber, fill the buffer chamber with ice cold 0.5X TBE buffer, attach the cover in the correct electrode orientation, and place the entire unit on a magnetic stirrer.

7. Connect the unit into a power supply and set the voltage at 25-30 V. Over a 5-10 minute period, check to be sure that the amperage is relatively stable and that no heating is evident. If the current changes significantly or heat is detected, it is likely that the buffer is incorrectly formulated.

8. Increase the voltage to 80 V and run for 1 hour.

9. Dismantle the apparatus and carefully clean the nylon membrane with buffer taking care to remove any adhering gel particles.

10. Bake the membrane in a vacuum oven at 80°C for 1-2 hours.

11. Place the membrane between two pieces of filter paper and store until ready for hybridization with the biotinylated probe.

Method, continued

Hybridization

Membranes or filters from Southern transfers are prepared for the purpose of detecting particular segments of DNA. This is done by hybridizing the membrane with a labeled probe complementary to the sequence to be detected. The labeling of the probe permits detection of the complementary strand of DNA to which it is bound on the membrane.

1. Soak the membrane from the Southern transfer in 2x SCC until completely hydrated (5 min.).

2. Dilute 10 ml of 2x prehybridization solution with an equal volume of ddH_20. Denature the prehybridization solution by heating in a boiling water bath for 10 minutes. Chill on ice for 10 minutes.

3. Pour the denatured prehybridization solution into a sealable plastic bag, insert the membrane, seal the bag and incubate at 42°C for 15 minutes. Open the bag and carefully pour the prehybridization solution into an appropriate container (this solution can be reused, so save it at -20°C).

   The prehybridization solution contains, among other ingredients, salmon sperm DNA. This DNA is nonhomologous with the prokaryotic DNA used in the exercise and will not interfere with the hybridization. The purpose is to block non-specific DNA binding sites on the membrane. If these sites are occupied by the salmon sperm DNA, the probe DNA can bind only where base-pair complement sequences occur. This procedure is necessary to prevent high background during the subsequent detection step.

4. Dilute 10 ml of 2x hybridization solution with an equal volume of ddH_20. Add an amount of probe DNA solution such that approximately 10 µg of probe DNA are used. This will give a probe concentration of 0.5 µg/ml (500 ng/ml) which is at the upper limit of recommended concentrations for biotinylated probes (100-500 ng/ml is the recommended range).

5. Denature by heating for 10 minutes in a boiling water bath and immediately chill on ice for 10 minutes. Add this to the bag containing the membrane, seal the bag and incubate at 42°C overnight.

6. Carefully pour off the hybridization solution in an appropriate container and save the solution for future use.

Method, continued

7. Wash the membrane in 200-300 ml 2x SSC, 0.1% SDS for 3 minutes at room temperature. Repeat.

8. Wash the membrane in 200-300 ml 0.2x SSC, 0.1% SDS for 3 minutes. Repeat.

9. Wash the membrane in 200-300 ml 0.15x SSC, 0.1% SDS for 5 minutes. Repeat.

10. Rinse membrane in 2x SSC at room temperature and proceed with the color detection. Note: the membrane may be dried and stored overnight in 2x SSC before starting the color detection.

Color Detection

The hybridized membrane will be incubated with a Streptavidin-Alkaline Phosphatase (SAAP) conjugate and, due to the strong affinity of streptavidin for biotin, this results in localization of the conjugate at only those membrane sites where probe/target complements are found. The sites can be detected, in turn, by a reaction between BCIP (5-bromo-4-chloro-3-indolyl phosphate) and the alkaline phosphatase component of the conjugate. In the reaction, alkaline phosphatase catalyzes the hydrolysis of phosphate from BCIP. BCIP then reacts with NBT (nitro blue tetrazolium) producing a blue precipitate at the site of the reaction.

1. Soak the membrane from the Southern transfer in 2x SSC until completely hydrated (5 min).

2. Block nonspecific protein binding sites on the filter by soaking the filter for 30 minutes in 5 ml Buffer B. This can be done with small filters using a petri plate or a pipet tip lid. The bovine serum albumin is used to block nonspecific protein binding sites on the membrane. This prevents generalized binding of Streptavidin-Alkaline phosphatase and reduces background during detection.

3. Add 5 ml of Streptavidin-Alkaline Phosphatase (SAAP) conjugate and incubate for 25 minutes.

4. Remove unbound Streptavidin-Alkaline Phosphatase conjugate by washing 3 times with 50-100 ml of Buffer A, 10 minutes for each wash.

Method, continued

5. Wash once with Buffer C for 5 minutes.

6. Prepare NBT/BCIP dye solutions:

   - Carbonate Buffer: Dissolve 8.4 g $NaHCO_3$ and 0.203 g $MgCl_2$ 6H20, in 800 ml ddH_20, adjust pH to 9.8 and dilute to 1 liter.

   - NBT Stock: Dissolve 30 mg of NBT in 0.7 ml dimethylformamide, 0.3 ml ddH_20.

   - BCIP Stock: Dissolve 15 mg of BCIP in 1 ml dimethylformamide.

7. For the detection, mix 1 ml of NBT stock and 1 ml of BCIP stock in 100 ml of the carbonate buffer.

8. Incubate the filter in this solution for 30 minutes - 3 hours in the dark.

9. Terminate the reaction by washing in 1 mM EDTA. Allow to dry away from light.

Results

The *Hind* III pattern of DNA fragments should become visible as blue bands within the first thirty minutes. The incubation can be extended to as long as three hours if bands are weak.

UNIT 6

Separation Technology

UNIT 6: SEPARATION TECHNOLOGY

MODULE 21: GEL FILTRATION CHROMATOGRAPHY

Barbara Crutch Jones

* Introduction
* Safety Guidelines
* Experimental Outline
* Materials
* Method
* Results

Introduction

The basic components of gel filtration chromatography are the matrix, column, and the elution buffer. The function of the matrix is to perform the separation. Separation is accomplished by pore sizes and internal channels available in the gel comprising the matrix. The gels are made of cross-linked polymers. The latter contributes to the fractionation range of each matrix and is referred to as the exclusion limit for each gel. A typical separation may appear as following:

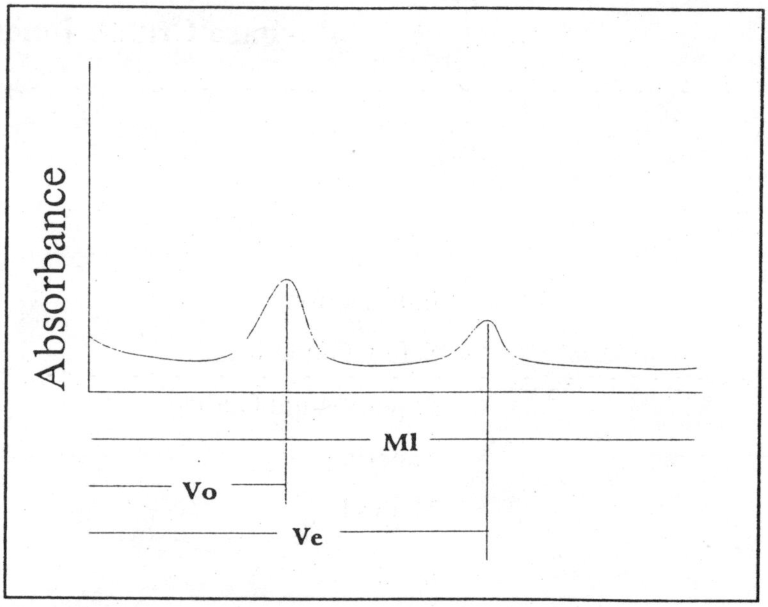

Figure 1: Void (V_o) and Effluent (V_e) Volume

The spectrum of molecular weights that the matrix is capable of separating is called the fractionation range. For example, consider a matrix that has a fractionation range of 1000 daltons to 100,000 daltons. Molecular weights of 1000 daltons or less will not be separated from each other, since they will penetrate the beads completely. These molecules take the maximum volume of the which is equal to one bed volume. Molecules having molecular weights in the range of 1000 to 100,000 daltons will enter the beads with varying efficiencies and be partially or completely separated from one another. Molecules having molecular weights greater than 100,000 daltons will not enter the beads. They will be eluted in the void volume. Thus, in regards to this example, any molecules with molecular weights greater than 100,000 daltons will elute at the same time, since they will not be sieved by the matrix. Partially or completely separated molecules eluted from the column are called peaks. Peaks consist of increasing and decreasing concentration gradients of molecules. An illustration is shown below.

Introduction, continued

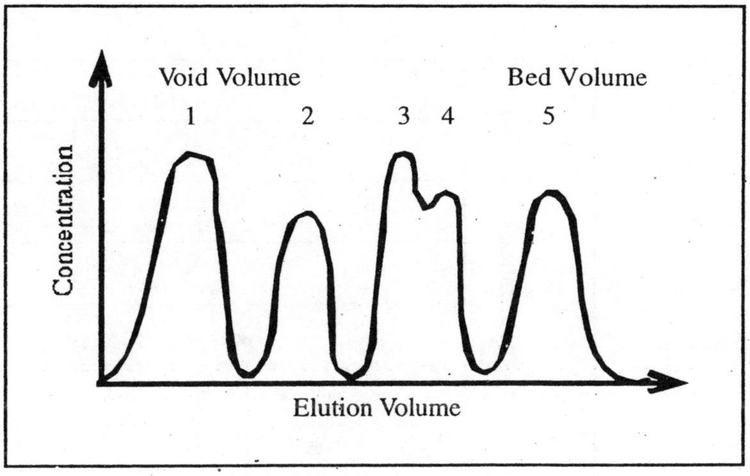

Figure 2

The column is a tube with a frit (an elution spout) fitted at the bottom. This frit is a membrane or porous disk that supports and retains the matrix in the column, but allows water and dissolved solutes to pass. The length of the column affects the resolutions of separation. Column lengths of 20 to 30 centimeters are frequently used. The diameter of the column as well as its length influence the kind of resolutions one may obtain during the chromatographic process. The elution buffer (the mobile phase of the chromatography) flows through the matrix and out of the column. The function of the elution buffer is to provide a means for developing the matrix with the applied sample contained in the column. This means that molecules in the sample are carried by the flow of a buffer into the matrix where they are gradually separated. In general, the larger the molecule the more difficult for it to pass through the pores and penetrate the beads. The shape of the molecule is also a critical factor in the separation. Thus, larger molecules tend to spend time flowing around and between the beads of the matrix, while smaller molecules tend to spend more time in the pores and maze of the channels of the matrix's beads. Consequently, the larger, higher molecular weight molecules are eluted from the column before, smaller molecules. Larger molecules take the faster and more direct path that involves less time in the column.

One can determine, graphically, the void volume (V_o) and effluent volume (V_e) for the separation process.

Introduction, continued

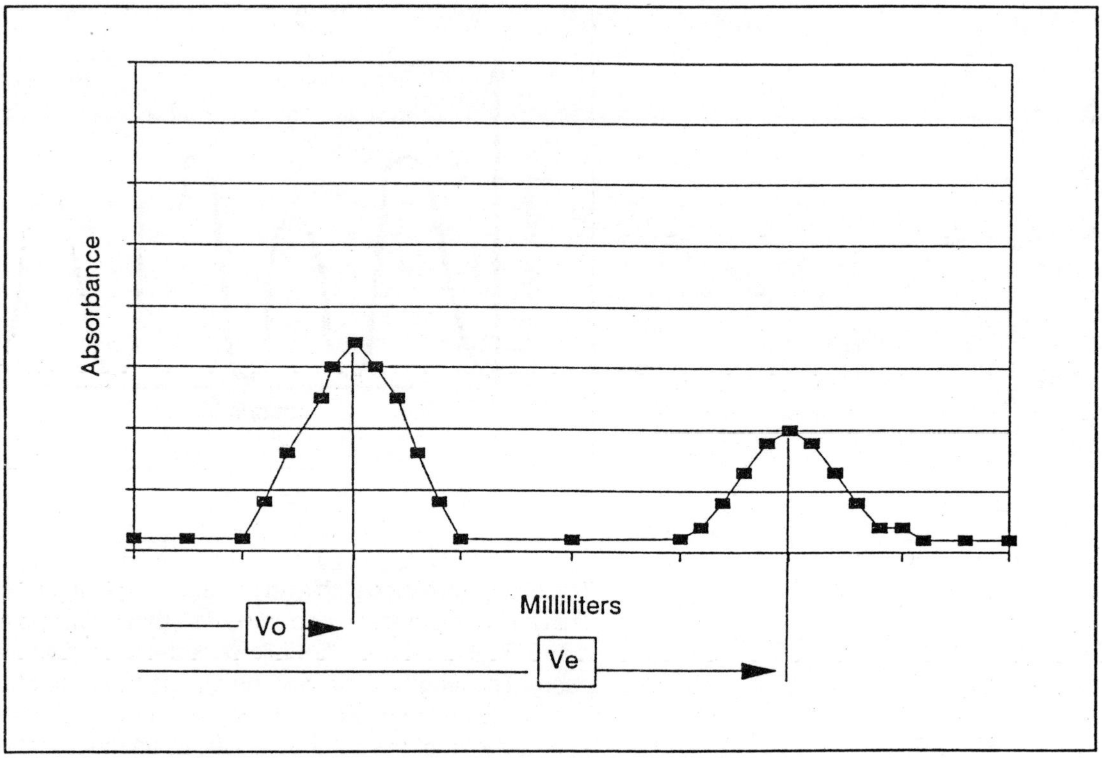

Figure 3: Void (V_o) and Effluent (V_e) Volume Determination

The void volume is the volume of the space surrounding and outside the particles of gel. It is usually determined by measuring the volume necessary to elute a solute that is excluded from the pores of the gel. A commonly used solute is blue dextran dye. This is a high molecular weight polysaccharide dye complex. The elution volume is the volume of solvent necessary to elute a solute from the time the solute enters the gel bed to the time it begins to emerge at the bottom of the column. The V_e is measured as the volume of solvent that has flowed through the column when the leading side of the solute peak is extrapolated to the base line of an elution profile. The total volume (V_t) is the total volume occupied by the packed gel bed. This is obtained most easily by a water calibration of column prior to packing it.

Introduction, continued

A plot of elution volume versus log of molecular weights for several reference substances of known molecular weights (a calibration curve) provides an excellent analytical tool for determining the molecular mass of a test molecule.

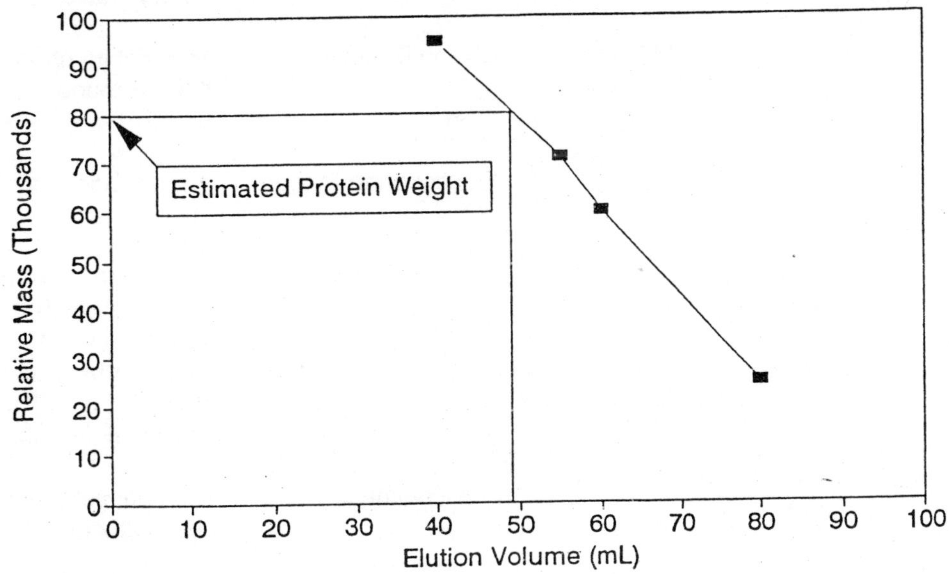

Figure 4: Molecular Weight Estimation of Proteins by Gel Filtration

For example, an elution volume of 49 ml will extrapolate to a molecular mass value of 8.0×10^4. One can also determine for each separated component analyzed a Pseudo Partition Coefficient (K_{av}). This Kav is comparable to Rf for Thin Layer Chromatography and Paper Chromatography.

$$K_{av} - \frac{(V_e - V_o)}{(V_t - V_o)}$$

Introduction, continued

Glossary of Terms:

Matrix	The solid support (gel) or the material in the column that actually performs the separation. It is the stationary phase of chromatography.
Column	A tube with a frit and elution spout or outflow valve fitted at the bottom.
Elution Buffer	The mobile phase of chromatography. It flows through the matrix and out the column.
Fractionation Range	An indication of the molecular size of a protein expected to elute at an elution volume equal to the bed volume (number on the left) and the molecular size of a protein expected to be totally excluded from the column and to elute at the void volume (the number on the right). Example: the fractionation range for Sephadex G-100 is (4,000 - 1 x 10^5).
Bed Volume	The total volume occupied by the packed gel, matrix or solid support.
Void Volume	The space surrounding and outside the particles of gel. It is determined by measuring the volume necessary to elute a solute that is totally excluded from the pores of the gel or matrix.
Exclusion Limit	The molecular weight of the smallest peptide or globular protein that will not enter the gel pore.

In this particular experiment, you will separate plasmid DNA which has been mixed with RNA. The plasmid is a circular DNA molecule with a molecular weight of approximately 4,000 base pairs. The RNA contaminants consist of degraded fragments of mRNA, rRNA, and tRNA with molecular weights in the range of 100 to several hundred base pairs. The samples also contain two dyes of different molecular weights which will aid in monitoring the chromatographic process and in locating the RNA and DNA.

Safety Guidelines

Always dress appropriately when performing any laboratory experiment. Wear gloves, an apron and safety glasses.

Experimental Outline

The objective of this experiment is to purify plasmid DNA that contains RNA using Gel Filtration Chromatography.

Pack Column
↓
Conduct Chromatography
↓
Conduct Electrophoresis

Materials

This module is based on EDVOTEK experiment #204. The following components are included:

Sample containing RNA, DNA and dyes
Slurry (matrix in buffer)
Concentrated Elution Buffer
Tube 10x Gel Loading Solution
Tube Practice Gel Loading Solution
Tris-acetate-EDTA electrophoresis buffer
Methylene Blue Plus™ or Ethidium Bromide stain
Columns with reservoir funnels
Valves
1 ml pipet
100 ml graduated cylinder (packaging for samples)
Transfer pipets

Requirements

Horizontal gel electrophoresis apparatus
D.C. power supply
DNA visualization system (white light for methylene blue optimal, U.V. transilluminator for ethidium bromide)
Automatic micropipets with tips
5 ring stands with clamps
1 ml and 10 ml pipets
Microtest tubes or glass test tubes 12x75 mm, 5 ml preferable, for column fractions)
Microtest (microcentrifuge) tubes or 1 ml test tubes (for preparing samples before electrophoresis)
500 - 1000 ml graduated cylinder
Small beakers, flasks, or tubes (10 - 25 ml, for slurry)
Beakers or flasks (50-100 ml, for column eluant and buffer)
Distilled water

Pre-lab Preparation

Preparation of Columns

1. Vertically mount the column on a ring stand, making sure it is straight.

2. Insert the reservoir funnel snugly into the top of the column.

3. Slide the wide opening of the valve onto the eluant spout at the bottom of the column. Make sure the valve fits securely.

4. Close the valve.

Preparation of Matrix Slurry

1. Thoroughly mix the slurry in Bottle B by inverting it several times.

2. Pipet (or pour) 9 ml into a small beaker or flask for each group.

Preparation of Elution Buffer

1. To 360 ml of distilled water, pour all of the contents of Tube C (concentrated elution buffer). Mix.

2. Dispense 50 ml of diluted buffer in a beaker for each lab group. Keep the extra buffer on hand in case of spills during the lab.

3. Pipet 1 ml of water in 5 microtest tubes. The level of water in the tube will be used as a reference guide for the equal collection of column effluent.

4. Before the column is packed, add 0.2 ml of buffer to each empty column. Make sure the buffer is at the bottom of the column.

Preparation of Sample Mixture

On the day of the Chromatography Experiment:

1. Add 2 ml of distilled water to vial A. Cap the vial and mix until all the freeze dried material is dissolved.

2. Dispense 0.3 ml to each of five tubes labeled "Sample".

Method, continued

Quick Reference:

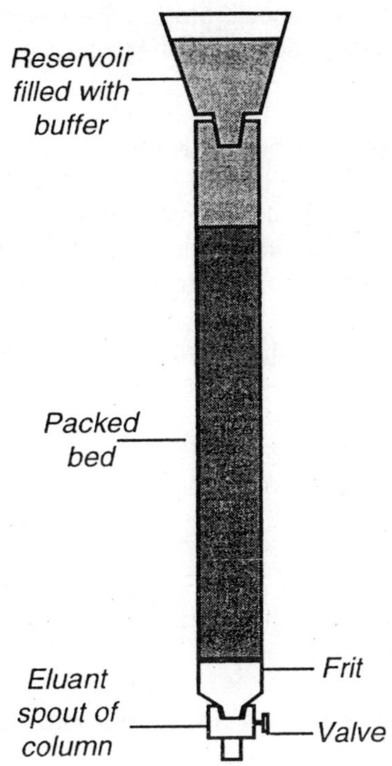

Figure 5

1. Packing the Column

   A. Immediately after mixing the slurry, pour all of it into the column, containing 0.2 ml of buffer, by letting it stream down the inside walls of the reservoir funnel. Slurry should be added slowly to the column. This will alleviate the development of air bubbles in the slurry.

   B. Add approximately 8.0 ml of buffer to the reservoir with a 10 ml pipet.

   C. Place a beaker under the column

   D. Open valve and allow buffer to flow through the column for approximately 10 minutes. This will cause packing of the matrix into the column. To ensure the removal of air pockets, tap gently on the column as it is compressing.

   E. Close the valve when the column is packed. The matrix is packed when it stops compressing. The bed volume should be 6 to 7 ml.

2. Chromatography

   A. Label 12 plastic micro test tubes as X, Y, Z and 1 to 9.

   B. Obtain tube with sample. Using a plastic pasteur pipet, put 4 drops of sample into Tube X. Cap and set aside. Return any remaining solution in the pipet to "sample" tube.

   C. Carefully remove all the buffer from above the bed (the column matrix) with a pasteur pipet. Insert the pipet through the reservoir funnel and into the column to remove the remaining buffer from the top of the matrix.

   D. Slowly load all the contents of the sample tube onto the top of the bed with a pasteur pipet. Allow the sample to run down the inside wall of the column. Try to minimize disturbance of the bed while loading.

   E. Place a beaker under the column.

   F. Open valve. The sample will slowly enter the bed. When sample has completely entered the bed, close the valve.

   G. Carefully add buffer to cover the bed using a pipet in the same fashion as the sample was applied. Continue adding buffer until the funnel is almost full.

Method, continued

H. Hold test tube #1 directly under column. Open the valve.

I. Using the micro test tube containing 1.0 ml of water as a guide, collect 1.0 ml of column effluent in each of the tubes (1-8). The dyes should gradually separate in the column. Periodically, add fresh buffer to the reservoir to keep it full.

J. After all eight tubes have been collected, close the valve.

K. Identify tube with the greatest amount of blue dye eluted from column. With a pasteur pipet, transfer nine drops (100 μl) from this tube into a micro test tube labeled "Y".

L. Identify tube with greatest amount of orange dye eluted from column. Transfer nine drops (100 μl) from this tube into a microtest tube labeled "Z". If time does not permit you to continue, this is a good place to stop the experiment. Tubes X, Y and Z are used for electrophoresis.

3. Agarose Gel Electrophoresis

A. Refer to Module 1, Unit 1 for guidelines for agarose gel electrophoresis procedures.

B. Load prepared samples and electrophorese.

C. Stain gel for visualization and analysis.

Results

The tubes containing the greatest amounts of each of the separated components of the sample mixture represents the chromatographic peaks. The dye, blue dextran, is associated in size with DNA, the larger molecule; therefore, DNA is expected to elute first from the column. DNA should be followed by RNA fractions which are associated in size with the dye orange G. Thus, the tube containing the greatest amount of blue dye will contain the plasmid DNA, while the tube containing the greatest amount of orange dye will contain the RNA fractions. These results will be confirmed upon analysis of samples by agarose gel electrophoresis.

Students may collect and report experimental data in the following manner.

I. Peak#	Elution Volume Value (ml)	Estimated K_{av}
1.	___________	___________
2.	___________	___________
3.	___________	___________
4.	___________	___________
Etc.	___________	___________
II. V_o Value (ml)	___________	
V_e Value (ml)	___________	
V_t Value (ml)	___________	

III. Draw an idealized schematic of electrophoresis results.

UNIT 6: SEPARATION TECHNOLOGY

MODULE 22: ION EXCHANGE CHROMATOGRAPHY

Richard E. Echols

* Introduction
* Experimental Outline
* Materials
* Pre-lab Preparation
* Method
* Results

Introduction

Most biological compounds are positive or negative charged when exposed to pHs in the range of 2-10. When the pH is varied, the net charge can change from zero to the opposite charge. In ion exchange chromatography, the solid support (adsorbent) contains either a permanent positive or negative charge. They are called cation and anion exchanger respectively. The permanent charge on the exchanger is attracted to the opposite charge on the molecules. The separation of compounds is based on an equilibrium between the molecules-exchanger and elution solvent. This equilibrium can be shifted gradually by changing the ionic strength or pH of the eluting buffer, thereby weakening the electrostatic forces and deadsorbing the molecules from the exchanger. This allows the separation of molecules with small differences in net charges.

The solid support is usually a synthetic resin (cross-linked polystyrene) or cellulose derivative covalently bonded to the desired functional group to create a weak or strong ion exchanger. A weak cation exchanger's functional group is a carboxylic acid and strong exchanger is sulfonate, whereas the anion exchangers are derivatives of either secondary or tertiary amines.

$$-CH_2COO^-, \quad R\text{-}SO_3H, \quad -CH_2\overset{+}{N}HR_2 \quad -CH_2\overset{+}{N}R_3$$

Carboxymethylcellulose (CM-cellulose) has the $-CH_2OH$ groups of cellulose derivatized to $-CH_2OCH_2COOH$, and the corresponding cation exchanger is substituted with $-CH_2OCH_2CH_2N(CH_2CH_3)_2$ (DEAE-cellulose). The exchanger is supplied from the manufacturer with a counter ion, which can be Na^+, H^+, Cl^- etc. The capacity of the exchanger is determined by the number of meq/ml of a standard material that can be adsorbed. Highly cross-linked and large capacity resins can be used for small molecules. The resin acts as a molecular sieve, which can block large molecules from entering the interior of the resin beads. In the case of cellulose, there is some limit to the number of substitutions that can be made per unit of cellulose. If it is too highly substituted, the support will become water soluble. Celluloses are the preferred supports for large biologically active proteins because they do not denature (deactivate) the protein as readily as resins.

The adsorption and separation are based on the differences between electrostatic interaction of the molecules and support. The following example demonstrates the exchange principle.

Introduction, continued

Support-COO^-Na^+

↓

+ NH_3^+-molecule (1)
+ $(NH_3)^+_2$-molecule (2)

Support-COO^- $^+NH_3$-molecule (1)
Support-COO^- $(+NH_3)_2$-molecule (2)

↓ first elution with NaCl

Support-COO^- $(+NH_3)_2$-molecule (2)

NH_3^+-molecule (1) (fraction 1 eluted)

↓ continue elution

Support-COO^-Na^+ +

(NH_3) molecule (1) (fraction 2)

Molecule 2 should have a greater attraction for the support than molecule 1. By changing the ionic strength or pH, the equilibrium point for molecule 1 should be attained before that of molecule 2.

Ion exchange chromatography can be used to separate both small molecules, such as amino acids and large ones like proteins, RNA and DNA.

Experimental Outline

A. CM-Sephadex
packed column

↓ add 0.2 ml of dye soln.

B. packed column + attached dye

↓

* C. Column + separated dye (1-in. separation)

↓

**D. blue dextran + DPN-glycine + Cyto. C

*C measure the volume of buffer used and distance traveled by components

**D measure the total volume of buffer need to completely elude the components.

Materials

CM-Sephadex (G-50) prepared in 0.1M potassium acetate buffer pH 6.0
0.1M Potassium acetate buffer, pH 6.0 (Buffer A)
1.0 Potassium acetate buffer, pH 6.0 (Buffer B)
Dye Mixture: Blue dextran (1 mg/ml), cytochrome c (2 mg/ml), and DNP-glycine (1 mg/ml)
10 ml chromatography column
Pasteur pipette with dropper bulb can substitute for the chromatography column
Rubber tube and pinch clamp

CM-Sephadex, is a weak anion exchanger ($RCOO^-$). When equilibrated in buffer at pH 6, it becomes $RCOO^-K^+$, and RCOOH when equilibrated in 0.1 M HCl. Blue dextran is a large nonionic polysaccharide, Molecular weight > 500,000, with a blue color. Cytochrome c's pI = 10.7 (pH where the net charge = 0), and has a net positive charge (cation) at the pHs of the experimental conditions. Cytochrome c has a molecular weight of 12,400. DNP-glycine is in the anionic form above pH of 3.0, has a molecular weight of 241 and a yellow color.

Pre-lab Preparation

Suspend the CM-Sephadex G-50 (30 gm) in 500 ml of H_2O and vacuum filter (Buchner funnel). Resuspend the CM-Sephadex in 0.1N HCl (83 ml conc. HCl into 920 ml H_2O), vacuum filter and rinse with 3-4 volume of H_2O (vacuum filter). Resuspend in 2 vol cf 1.0 M potassium acetate, pH 6.0. and vacuum filter. Repeat this process a second time (final cake). Resuspend the CM-Sephadex cake in 2 vol of the 1.0 M potassium acetate buffer for column packing.

Repeat (or take half of final cake) and resuspend in 2 vols of 0.1 M potassium acetate, pH 6.0. Vacuum filter and repeat the process a second time. Resuspend the cake in 1 vol 0.1M potassium acetate.

Buffers:

1.0M potassium acetate, pH 6.0. Dissolve 98 gm of potassium acetate in 950 ml of H_2O in a beaker and adjust to pH 6.0 with glacial acetic acid.

0.1 M potassium acetate pH 6.0 Take 100 ml of 1.0 M potassium acetate buffer and dilute to 1000 ml.

Sample:

In a sample vial add 1 mg of blue dextran, 2 mg cytochrome c and 2 mg of DNP-glycine and dilute to approximately 5 ml.

Method

1. Set up two columns. Place a piece of glass wool in the restriction and label each column.

2. Attach a piece of rubber tube and clamp to each column. Close the clamp.

3. Fill a column with each of the CM-sephadex buffer solutions (buffer A or B). Add the slurries using a pipette and bulb. Open the pinch clamp and allow the slurry to settle until resin bed is 4-5 in. Do not allow the column to run dry. Add the proper buffers (as used for packing columns) dropwise to maintain the liquid level just above the resin bed. Close the clamps.

4. Add approximately 0.2 ml of the solution to be separated. Form a narrow band at the top of the resin. Any of the solution that may adhere to side of column above the resin rinse with a few drops of the buffer on to the resin (gradually open the clamp to allow the solution to move onto the resin bed).

Method, continued

5. Place a 10 ml graduated cylinder under each column.
6. Open the clamp, and fill the column with buffer.
7. Allow the buffer to flow through the columns until a clear separation of at least an inch has taken place. Measure the distance the bands has travel in centimeters from the top of the resin. Measure the amount of buffer collected from each column.
8. Relate the distance travel by the band to the amount of buffer collected.
9. Measure the volume of buffer required to elute each band (collect the bands together).

Results

Table for Results After One Inch Separation (Approx)

	0.1 M (Buffer A)			1.0 M (Buffer B)		
	A	B	C	A	B	C
Vol Buffer						
distance (cm)						
R_f*						
R_f						

$R_f = \dfrac{\qquad}{\text{volume collected}}$ or $R_f^* = \dfrac{\qquad}{\text{total volume of elution*}}$

* total elution volume to remove the component

A = dextran B = DNP-glycine C = cytochrome c

UNIT 6: SEPARATION TECHNOLOGY

MODULE 23: AFFINITY CHROMATOGRAPHY

Baldwin King

* Introduction
* Safety Guidelines
* Experimental Outline
* Materials
* Method
* Results

Introduction

In affinity chromatography, the material to be separated or purified (S) binds to a ligand (L) immobilized on an insoluble matrix (M) such as agarose. The technique represents a powerful and efficient method for separating proteins and nucleic acids. The general protocol is to immobilize the ligand by covalently binding it to a suitable matrix. The substance of interest is then bound with high specificity to the ligand to form a complex while the impurities pass through the column. The substance S is finally eluted from the column by an eluent which has an even higher binding capacity for S than the immobilized ligand, L.

Several reactions are used to couple the ligand to the solid matrix depending on the nature of the ligand and the type of matrix. In the present application known as immobilized metal affinity chromatography (IMAC), a chelating ligand, iminodiacetate (IDA) is coupled to epoxy-activated agarose. A transition metal such as Cu^{2+} is then bound to the iminodiacetate. Other metal ions used are Zn^{2+}, Fe^{3+} and Ni^{2+} and chelating ligands include tris (carboxyethyl) ethylenediamine (TED). The IDA ligand, tridentate, occupies three coordinating positions around the central metal ion thus leaving three other coordinating positions for electron donors like proteins and nucleic acids containing N, O or S atoms on their surfaces.

When a mixture of (serum) proteins is added to the column, some proteins, especially those containing histidine residues, are preferentially bound to the metal. The degree of binding depends on the type of protein, the pH and ionic strength of the surroundings. The pH should be such that the protein is not too highly protonated. Thus buffers with pH close to neutral and weakly coordinating are frequently used during the adsorption stage. Acetate (pH 5-6) and phosphate (pH 8-9) are common choices.

Unbound proteins pass through the column fairly rapidly and so a partial separation is effected. The bound proteins can then either be eluted as a group or individually by proper choice of eluting conditions. The most obvious choice is to lower the pH so that the protein is protonated and therefore released from the metal-protein complex. However care must be taken not to denature the protein at these low pH. Another common method of desorption is to use a ligand in the eluent that has a stronger affinity for the metal than the protein. Tris (Hydroxymethyl) aminomethane (tris) buffers and imidazole are good choices.

A 'model' protein system has been chosen consisting of a mixture of rabbit serum IgG and rabbit serum albumin (RSA). The affinity column is IDA-agarose with bound Cu^{+2}. The experiment will serve to demonstrate the usefulness of the technique in separating the two proteins. The effect of metal type on the efficiency of separation could also be examined.

Safety Guidelines

1. Safety glasses should be worn.
2. When handling the proteins, gloves should be worn to avoid contamination.
3. All chemicals and biologicals should be disposed of in labeled containers.

Experimental Outline

1. Bind metal to IDA - agarose column
2. Apply protein mixture to column
3. Isolate first protein fraction by washing with buffer A
4. Measure absorbance of first set of fractions at 280 nm
5. Isolate second protein fraction by eluting with buffer B
6. Measure absorbance of second set of fractions at 280 nm
7. Identify proteins by immunoelectrophoresis or radial immuno-diffusion (optional)

Materials

5 ml rabbit serum albumin (RSA) solution at 1 mg/ml

2 ml of IgG solution at 0.5 mg/ml

Protein mixture: 1 ml of a mixture of rabbit serum IgG and rabbit serum albumin (0.5 ml each of above solutions)

100 ml of 50 mM sodium acetate and 100 mM NaCl, pH 5.5 (buffer A)

100 ml of 100 mM Tris-Cl buffer, pH 8.0 (buffer B)

10 ml of 50 mM Cu^{2+} solution (as copper (II) sulfate in water)

100 ml of 50 mM EDTA and 0.5 M NaCl solution, pH 7.0

2 prepacked 0.8- x 5-cm IDA-agarose columns (Sigma IDA-5)

UV-VIS spectrometer

1 80-tube fraction collector (optional)

1 Peristaltic pump (optional)

6 Pasteur pipets and bulb

Electrophoresis or Immunoelectrophoresis apparatus (optional)

Method

1. Clamp the prepacked IDA-agarose column in a ring stand.
2. Drain the solution from the column to a level just slightly above the top of the gel. (Do not allow column to go dry).
3. Wash the column twice with 2 ml portions of Buffer A. Discard the wash.
4. Add 1 ml of the 50 mM Cu^{2+} solution to the column with a Pasteur pipet and allow it to enter the gel by slowly opening the outlet.
5. Fill the column with buffer A (or pump it) and continue to wash the column until all excess Cu^{2+} has been eluted (about 15 ml).

Method, continued

6. Attach column outlet to a fraction collector or alternatively collect fractions by hand

7. Add 1 ml of the protein mixture (RSA and IgG mixture) to the column and allow it to enter the gel.

8. Fill the column with buffer A (or pump it) and begin to elute. Collect 2 ml- fractions .

9. Measure the absorbance at 280 nm of your fractions from the start of the elution process and continue to do so until the absorbance returns to zero (about 10 ml; use buffer A as reference).

10. Now fill the column with buffer B and elute as you collect 2 ml-fractions again.

11. Monitor the absorbance at 280 nm of your new set of fractions until the absorbance returns to zero (about 20 ml; use buffer B as reference).

12. Pool all the fractions eluted with buffer A that contain protein and likewise pool all the fractions eluted with buffer B that contain protein.

13. Identify the protein (after concentration by your instructor) in each sample by PAGE electrophoresis or immunoelectrophoresis.

Results

1. Set up the following tables to record your results.

Table I		Table II	
Fraction #	A_{280}	Fraction #	A_{280}
1		11	
2		12	
3		13	
4		14	
5		15	
6		16	
7		17	
8		18	
9		19	
10		20	

Results, continued

2. Plot the data on 2 separate graphs with A_{280} as the ordinate and fraction # as the abscissa.

3. Discuss briefly the results in terms of the efficiency of the separation, the identity of the protein fractions, the choice of chelating metal, the choice of buffer systems and any recommendations for improving the method.

UNIT 6: SEPARATION TECHNOLOGY

MODULE 24: PURIFICATION OF WHEAT GERM PHOSPHATASE

Barbara Crutch Jones

* Introduction
* Safety Guidelines
* Experimental Outline
* Materials
* Pre-lab Preparation
* Method
* Results

Introduction

Major activities of the biochemist are the isolation and purification of biological molecules such as proteins, nucleic acids, and carbohydrates. The methods include organic solvents precipitation, salting out, and selective denaturation.

The first recognized cell - free enzyme preparation was made from yeast by E. Buchner in 1897; additional work on isolated and purified protein was conducted by Menten in 1913. Since the work of Menten, others have successfully isolated proteins in pure form for study. One method which appears to be used successfully for many of these isolations is Salting Out. This technique depends mainly on the hydrophobic character of the protein. Water solvates added salt ions decreasing the solvation of the protein molecules. This decrease exposes hydrophobic areas of the protein structure which interact with each other to give aggregates that precipitate. The procedure involves dissolving the salt into the solution containing the protein, or the addition of a saturated solution of the salt to the protein in solution form. The most commonly used salt is ammonium sulfate which has high water solubility. Its solubility is relatively temperature independent, and varies very little between 5 and 30°C. The optimum concentration of ammonium sulfate needed to precipitate a protein can be easily determined.

In this particular experiment, phosphatase is the enzyme of interest. It is isolated from wheat germ because this is an excellent inexpensive source of the enzyme. The presence of protein is detected by the biuret reagent. This protein forms a color complex with the copper constituent of the biuret. The color complex can be determined spectrophotometrically. If a protein in known varying concentrations is prepared and the absorbance determined, a standard concentration curve can be constructed. From such a curve, one can easily determine the concentration of an unknown protein, provided an absorbance value is obtained. This concept is made possible through Beer's Law, which states that absorbance is directly proportional to the concentration. To determine the activity of the phosphatase enzyme, the substrate p-nitrophenol phosphate is used. This substrate, in alkaline solution following hydrolytic cleavage, produces the p-nitrophenolate ion which exhibits a characteristic yellow color. The intensity of this color is used to detect and determine the activity of the enzyme which catalyzes the hydrolytic cleavage of the substrate.

Safety Guidelines

Always wear the proper attire for laboratory exercises. The attire should include gloves, safety glasses, and apron. Be very careful with your pipetting. Always use a pipette bulb. Never use your mouth. Be careful with the ammonium sulfate reagent. This is a very corrosive substance. If it is spilled, clean the area immediately with cold water. Follow the instructions given for the proper use of the centrifuge available. Be certain to balance all centrifuge tubes during centrifugation. Fill the centrifuge tubes to only 2/3 of their capacity. Make sure that all protein solutions are kept at 4°C or frozen until they are ready to be assayed. All standard protein solutions should remain frozen, and thawed just before they are to be used. The p-nitrophenylphosphate substrate should be prepared on the day it is to be used. Make certain that you are familiar with precautions associated with spectrophotometry. Your instructor will supply you with these.

Experimental Outline

DAY 1. Extraction of acid phosphatase through liquid fraction.

Water extraction of wheat germ

Salt fractionation of manganese chloride

Salt fractionation of ammonium sulfate-35%

Precipitation of phosphatase by ammonium sulfate -57% and heat treatment

Water extraction of phosphatase from residue fraction

DAY 2. Determination of protein concentration and phosphatase activity (phosphatase assay)

Development of protein standard curve

Determination of protein concentration on enzyme fractions

Determination of phosphatase activity on enzyme fractions - Phosphatase assay

Materials

- Wheat germ (approximately 12.5 grams)
- Beakers: 150 ml (3), 250 ml (2)
- Styrofoam box with crushed ice
- Refrigerated centrifuge
- Balance
- Graduate cylinders: 25 ml, 100 ml (5), and 500 ml (1)
- Magnetic stirrers and stirring bars
- Pipetes: 1.0 ml (5), 5.0 ml (5),10.0 ml (4), 100 μl
- Test tubes (20)
- Thermometer
- Water bath, adjusted to 60 to 65°C
- Rounded end glass rods
- Pasteur pipets
- Erlenmeyer flasks -150 ml
- Parafilm
- Refrigerator for storing solutions and insoluble fractions
- Test tube racks
- Large beaker (4 liter capacity) for soaking pipettes
- Freezer for storing enzyme fractions

Materials for Protein Determination

- Bovine Serum Albumin (BSA)
- Distilled water
- Biuret reagent
- Test tubes (12)
- Pipets - 1.0 ml (10), 5.0 ml (10), 100 μl and 8 tips
- Test tube rack
- Large beaker for soaking pipets
- Spectronic 20 and cuvets (2)

Materials for Phosphatase Assay

- Sodium acetate buffer (1.0M), pH 5.7
- Magnesium chloride (0.1M)
- p-nitrophenylphosphate (0.05M)
- Potassium hydroxide (0.5M)
- Distilled water
- Spectronic 20 and clean spectronic 20 tubes (cuvets)
- Parafilm
- Stopwatch
- Pipets
- Test tubes (8)
- Test tube rack
- Water bath
- Clinical centrifuge

Materials, continued

Dispense equipment (per student group)

- Beakers 150 ml, 250 ml
- Styrofoam box for crushed ice
- Refrigerated centrifuge
- Balance
- Graduate cylinders - 25 ml, 100 ml, and 500 ml
- Magnetic stirrers and stirring bar
- Pipets - 1.0 ml, 5.0 ml, 10.0 ml, 100 μl
- Test tubes
- Water bath (per five groups)
- Glass rods
- Thermometer
- Pasteur pipets and bulbs
- Test tube rack
- Spectronic 20 and cuvets
- Large 4 liter beaker for soaking pipets
- Desk top clinical centrifuge

Pre-lab Preparation

Dispense reagents

Bovine Serum Albumin (Frozen)
Copper sulfate (1.50 grams)
Sodium potassium tartrate (6.0 grams)
Sodium hydroxide (10 grams)
Distilled water (1.0 liter)

Preparation of solutions

Bovine Serum Albumin (1.0 mg/ml) - Dissolve 1.0 g of bovine serum albumin (Sigma, Fraction V) in about 80 ml of distilled water and dilute to a final volume of 100 ml. The BSA should be stored frozen in labeled containers and thawed only when needed for standard curve construction.

Biuret reagent - place 0.375 g of copper sulfate ($CuSO_4 \cdot 5H_2O$) and 1.5 g sodium potassium tartrate ($NaKC_4O_6 \cdot 4H_2O$) in a dry 250 ml volumetric flask. Add 125 ml of distilled water and dissolve the contents with shaking, then add enough distilled water to give a final volume of 250 milliliters. This should produce a solution of deep blue color. It can be stored in a plastic bottle at room temperature for an indefinite period.

Sodium acetate buffer (1.0M, pH = 5.7) - Dissolve 5.74 ml of glacial acetic acid in 80 ml of distilled water. Check the pH and adjust it to 5.7 by adding small volumes of 10.0 M NaOH (40 g NaOH dissolved in 100 ml of distilled water). Once the pH has been adjusted, add enough distilled water to give a final volume of 100 ml.

Magnesium chloride (0.1M) - dissolve 2.03 g of $MgCl_2 \cdot 6H_2O$ in distilled water to a final volume of 100 ml.

p-nitrophenylphosphate (0.05M) - dissolve 1.68 g of disodium p-nitrophenylphosphate in distilled water to a final volume of 100 ml. This solution should be prepared on the day it is to be used.

Potassium hydroxide (0.5M) - dissolve 5.62 g of KOH in distilled water to a volume of 100 milliliters.

Method

(Day 1)

I. Water extraction of the enzyme from the wheat germ

A. Transfer 12.5 grams of wheat germ to a 150 ml beaker surrounded by crushed ice in a styrofoam box.

B. Add 50 ml of cold distilled water, mix, and allow the mixture to stand for 30 minutes with occasional swirling. This will remove the acid phosphatase which is water soluble along with other water soluble substances.

C. Centrifuge at 6000 x g at 4°C for 10 minutes to separate the insoluble cellular material from the soluble protein.

D. Carefully decant the liquids of the centrifuge tubes after centrifugation. Combine all liquids in a chilled graduate cylinder. Record the volume. This represents (L-1). The residue at the bottom of the tubes (R-1) may be discarded. Save 1.0 ml of (L-1) for protein and enzyme analysis.

2. Salt fractionation of the enzyme with manganese chloride

A. Transfer (L-1) to a 150 ml beaker in an ice bath on top of a magnetic stirrer. Place in a stirring bar and stir the solution gently.

B. Add 1.0M $MnCl_2$ [2.0 ml/100 ml (L-1)] to the mixture. Make this addition very slowly, with stirring.

C. Centrifuge at 6000 x g at 4°C for 10.0 minutes. Combine the liquids in each centrifuge tube (L-II) and discard the residues (R-II).

D. Determine the total volume of (L-II) and save 1.0 ml of this liquid for enzyme and protein analysis.

3. Salt fractionation of the enzyme with 35% ammonium sulfate

A. Transfer the supernatant from above (L-II) to a 250 ml beaker in an ice bath on top of a magnetic stirrer. Place in a stirring bar and add 54 ml of cold saturated $(NH_4)_2SO_4$ (pH = 5.5) for every 100 ml of (L-II).

B. Ammonium sulfate should be added slowly from a pipette over a 20 to 25 minutes period. The stirring rate should be slow enough to alleviate any foaming in the solution.

Method, continued

C. Centrifuge at 6000 x g at 4°C for 10 minutes. Save the supernatants (L-III), measure the total volume, and set aside 1.0 ml of this fraction for protein and enzyme analysis.

4. Precipitation of the enzyme by 57% ammonium sulfate at 60°C.

A. Add cold saturated ammonium sulfate (pH= 5.5) in the same fashion as above. For every 100 ml of (L-III), add an additional 79 ml of saturated ammonium sulfate.

B. At the completion of the addition of ammonium sulfate, transfer the beaker to a hot water bath (60 to 65°C). With gently swirling, allow the temperature of the mixture to equilibrate to 60°C. Allow the mixture to remain at this temperature for 2 minutes, then transfer the mixture to an ice bath and cool the mixture to 5°C, rapidly.

C. Centrifuge at 6000 x g at 4°C for 10.0 minutes to collect the enzyme. At this step the enzyme should be in the insoluble residue .

D. Keep the residue (R-IV). Retain (L-IV) also until the presence of the phosphatase has been confirmed in (R-IV) via enzyme analysis.

5. Water extraction of the enzyme from (R-IV).

A. To each centrifuge tube containing the protein pellet residue (R-IV) add 10.0 ml of cold distilled water. Resuspend the residue by gently mixing with a glass rod to a uniform suspension.

B. Measure the total volume of the resuspended residue (R-IV). Add enough water to give a final volume equal to one third the volume of (L-II).

C. Centrifuge the mixture at 10,000 x g for 10.0 minutes. Keep (L-V), measure its total volume and retain 1.0 ml for enzyme and protein analysis. Discard the residue, (R-V).

D. Place all of the protein fractions in the freezer until the next laboratory meeting.

Method, continued

(Day 2)

1. Development of protein standard curve and determination of protein concentration.

   A. Bovine Serum Albumin (BSA) for a protein standard curve.

      1) Obtain a sample of (BSA) having a concentration of 10 mg/ml and prepare a series of 1.0 ml solutions. This series should have a range of protein concentration from zero to 10 mg/ml.

      2) To each sample add 4.0 ml of Biuret Reagent. Mix well and after 30 minutes at room temperature, read the absorbance of each sample at 540 nm.

      3) Record the absorbances and concentrations and prepare a standard curve by plotting.

   A_{540} versus protein concentration (mg/ml). Check with your instructor for the proper operations of the Spectronic 20.

   B. Determination of protein concentration in enzyme fractions.

      1) Set up test tubes (one for each sample that you wish to measure). Label these tubes and place 0.1 ml of each sample in its respective tube. (L-I), (L-II), (L-V), etc.

      2) Add 0.9 ml of distilled water to each sample.

      3) Add 4.0 ml of biuret reagent to each sample and let the samples stand at room temperature for thirty minutes.

      4) Determine the absorbance at 540 nm for each sample and, using the protein standard curve, determine the protein level (mg/ml) for each fraction.

      5) Enter information in the space provided in Table I.

Method, continued

2. Enzyme activity (phosphatase assay) - Acid phosphatase activity is determined by using the substrate p-nitrophenyl-phosphate. Two test tubes are set up for a single sample analysis. These two tubes should have the same composition of substrate, the same temperature, etc. However, to one tube the enzyme should be added, to the other (blank) an equal volume of water should be added. The reaction is allowed to proceed for 5 minutes and it is then terminated by the addition of potassium hydroxide. The contents of the tubes are compared spectrophotometrically. The amount of p-nitrophenol produced per minute is used as a measure of enzyme activity.

   A. Phosphatase assay

      1) Prepare two labeled test tubes (I and II) each containing the following reaction mixture:

         0.5 ml of 1.0 M Sodium Acetate buffer (pH = 5.7)
         0.5 ml of 0.1 M $MgCl_2$
         0.5 ml of 0.05M p-nitrophenylphosphate
         3.3 ml of distilled water

      2) Mix the contents of each tube and place them in a water bath (37°C). for 5 minutes.

      3) Add 0.2 ml of distilled water to tube I and 0.2 ml of each enzyme fractions to the properly labeled test tubes. Start a stopwatch and mix and return the tubes to the to the water bath. Tube I is the blank.

      4) After 5 minutes of reaction time, add 2.5 ml of 0.5M KOH to each tube. If the tubes appear turbid, centrifuge for 10 minutes and discard the precipitate. (Samples should be applied and reaction should be stopped so that each tube is allowed only five minutes of reaction time.)

      5) Using the contents of tube I as a blank, determine the absorbance of tube II, and remaining enzyme fractions at 405 nm. If the absorbance is greater than 1.0, the assay must be repeated using a more dilute solution for the enzyme. The reading at 405 nm should be converted to micromoles of p-nitrophenol formed per minute. The extinction coefficient for p-nitrophenol is 18.8 x 10^{-3} liter $mole^{-1}$. This

Method, continued

is true if measurements are made in a 1.0 cm length tube. Keep track of sample dilutions so that you can accurately calculate enzyme activity.

6) Determine the phosphatase activity (Units/ml) and total phosphatase activity (units) for all samples. Enter the data in Table 2. One unit of acid phosphatase is defined as the amount of enzyme that catalyzes the hydrolysis of 1.0 micromoles of p-nitrophenylphosphate per minute. Steps IIA 1-5 should be repeated for all the samples to be tested.

7) At this point, you will have completely salted out a protein. The enzyme may be further purified by solvent extraction and desalted by dialysis or chromatographically (see Unit 6, Module 25).

Results

Result Table for Steps I-A and B

Table 1. Acid Phosphatase Purification

Fraction	Vol (ml)	Sample (ml)	Protein (mg/ml)	Total Protein (mg)
L-I	39.0	0.1	44.7	1743.3
L-II				
L-III				
L-IV				
L-V				

The concentration for each fraction may be estimated from the standard protein concentration curve directly or it may be calculated by the following equation:

$$C_u = C_s\,(A_u/A_s)\,D$$

Where C_u = The concentration of the unknown solution
C_s = The concentration of the standard solution
A_u = The absorbance of the unknown solution
A_s = The absorbance of the standard solution
D = The dilution factor

Sample Calculation:

Using L-I as the sample, the standard Bovine Serum Albumin of 10 mg and the experimentally determined values of optical density (A_u = 0.255 ; A_s = 0.570) and D = 1.0, the calculation becomes:

C_u = 10.0 mg x 0.255 / 0.570
= 4.47 mg of protein in a test sample size of 0.1 ml.
C_u = 4.47 mg of Protein/ 0.1 ml
= 44.7 mg of protein

The total volume of the L-I fraction was 39.0 ml. Therefore, the total number of milligrams of protein may be computed as :
Total mg of Protein = 44.7 mg/ml x 39.0 ml
Total mg of Protein = 1743.3 mg

All subsequent fractions may be evaluated in the same manner.

Results, continued

Results Table for Steps II-A (1-5)

Table 2 Enzyme Activity for Acid Phosphatase Enzyme

Fraction	Activity (Units/ml)	Total Enzyme (Units)	Specific Activity (Units/ Mg of protein)
L-I	8.5	332.0	0.200
L-II			
L-III			
L-IV			
L-V			

Sample Calculation of Specific Activity:

Specific Activity = (Units/ml)/ (Mg of Protein/ml)
Units = micromoles of substrate hydrolyzed/ min-ml

Micromoles of substrate are obtained from Beer's Law:

A = absorbance of sample at designate wavelength, 405 nm in this case
a = the extinction coefficient for p-nitrophenol which is 18.8×10^{3} L/M-cm
b = the path length of light through the sample, 1.0 cm
c= the concentration of the sample, micromoles of p-nitrophenol formed per minute).

For L-I, absorbance was 0.160. Concentration computed as follows:

C = $0.160/(18.8 \times 10^{3}$ L/M-cm) x 1.0 cm
= 8.5 x 10-6 M/ L-5 min
= 1.7 x 10-6 M/L-min or 1.7 micromoles/min.

The sample size was 200 μl or 0.2 ml. This corresponds to 8.5μMoles/ml or 8.5 units/ml. The total units may be obtained from:

Total Units in L-I
= 8.5 units/ml x 39.0 ml = 332.0 units.

Substituting in the expression for the specific activity , one obtains:

Specific Activity = (Units/ml) / (mg of protein/ ml)
= (8.5 units/ml) / (44.7 mg protein/ml)
= 0.200 units/ mg of protein.

All subsequent fractions may be evaluated in the same manner.

UNIT 6: SEPARATION TECHNOLOGY

MODULE 25: CHROMATOGRAPHIC DESALTING OF PROTEINS

Barbara Crutch Jones

* Introduction
* Safety Guidelines
* Experimental Outline
* Materials
* Method
* Results

Introduction

Since the introduction of gel filtration by Porath and Flodin in the late 1950s, this technique of chromatography has been widely used to purify macromolecules.

The gel or matrix provides the stationary phase of this type of chromatography, while the mobile phase is the solution of molecules to be separated and the eluting solvent, which most frequently is water or a dilute buffer. The sample is applied to an appropriate sized column. If the molecules of the sample are too large for the pores of the gel of the column, they do not enter the gel, but move outside of the gel beads with the eluting solvent. The larger the molecules to be separated, the faster they move throughout the gel bed, leaving behind the smaller molecules which can enter the gel pores. The smaller molecules are retarded by this permeability and move at a slower rate through the gel bed. The differentiation in the rate of movement through the column provides a means of molecule separation and thus is a step in purification. Hence, molecules are eluted in order of decreasing molecular size as shown in Figure 1.

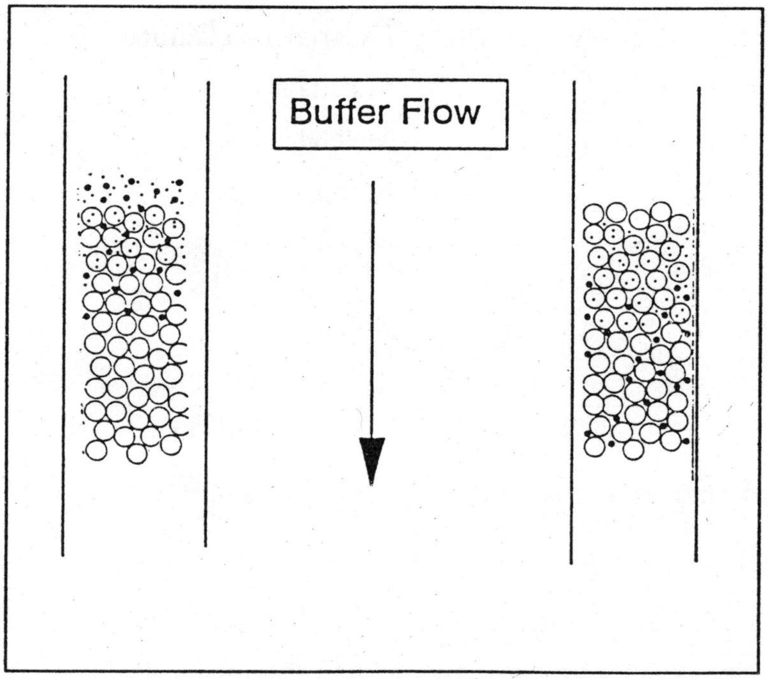

Figure 1 - Gel filtration chromatography. Open circles represent the porous gel molecules; large solid circles represent molecules too large to enter the gel through the pores, and smaller solid circles represent molecules capable of entering the gel pores.

Introduction, continued

Bed Volume	The total volume occupied by the packed gel, matrix or solid support. This is obtained most easily by a water calibration of the column before packing.
Buret	A graduated glass tube with a small aperture and stopcock for delivering measured quantities of liquid.
Column	A tube with a frit and elution spout or outflow valve fitted at the bottom.
Frit	A membrane or porous disk that supports and retains the matrix in the column, but allows water and dissolved solutes to pass through.
Matrix	The solid support or the material in the column containing elution solvent.
Slurry Pack	Matrix that is suspended in an elution solvent, properly swollen and poured into a column containing elution solvent.
Elution Solvent	The mobile phase of the chromatography. It flows through the matrix and out the column. The column with the matrix and applied sample is "developed" during the flow of this solvent.

Safety Guidelines

Always use protective covering for eyes, hands, and body when performing an experiment. Wear goggles, disposable gloves, and an apron while in the laboratory.

Be careful with Dextran Blue; avoid contact and inhalation.

Experimental Outline

1. Select the suitable chromatography matrix and elution solvent or buffer.
2. Equilibrating the matrix with buffer and packing the column. See the appendices for packing the column.
3. Setting up the equipment
4. Equilibrating the column with buffer
5. Applying the sample to the column
6. Eluting the column - the separation process
7. Analyzing the collected fractions for the amount of protein

Sometimes it is necessary to repeat steps 5 and 6 to enhance the separation process.

Day 1. Pack and equilibrate the column with buffer
Label fraction collecting tubes
Determine the column's void volume

Day 2. Apply the sample to the column
Elute the desalted protein
Analyze the eluted fractions for protein

Materials

A column

Ring stand and clamps

Gel filtration matrix (4 grams or dry Sephadex G-25)

Gel filtration buffer (Tris phosphate, 0.05M)

Gel filtration protein standard containing Blue Dextran
(These can be purchased from Sigma, Bio-Rad, Pharmacia or EDVOTEK)

90°C water bath

Desalting solution (0.03M NH_4OH or 0.05M Acetic Acid)

Lyophilization apparatus or a trapped vacuum pump

NaCl (10% w/v)

Buchner funnel

Carpenter's level

Powder funnel for a buffer reservoir

Small beaker for a gel reservoir

Spectronic 20 with at least four cuvettes

100-115 small test tubes

Biuret reagent

One filtration flask (25 ml)

Supply of pasteur pipets and bulbs

Approximately five grams of fine grid silica sand (clean)

One screw clamp

Tygon tubing

Glass rod and a piece of glass tubing

Paraffin paper - one sheet

One small test tube containing 1 ml water as a reference for collecting 1 ml fraction of effluents from the column

Flashlight

Aspirator

Method

Swelling the Gel

1. Piace 4 grams of dry Sephadex G-25 in gel filtration buffer (approximately 20 ml of buffer) and allow it to swell for approximately 5 hours in a beaker that is occasionally stirred gently with a glass rod in a water bath that is 90°C. Your instructor may swell the gel for you prior to the beginning of the laboratory period.

    a. Allow the gel to settle. Then aspirate or decant the fine gel particles that do not settle into the gel bed.

Method, continued

b. Suspend the settled gel in an equal volume of buffer to make a thick suspension (slurry). Pour the slurry into a filtration flask and degas the gel in order to remove trapped air. Allow the gel to reach the desired temperature.

Packing the Column

1. Mount the column vertically. Use the carpenter's level to ensure that the column is vertical.

2. If your column comes equipped with a frit at the outlet end, close the outlet valve and fill the column to 1/2 full with buffer. Check the outlet valve for leaks.

   If the glass wool is to be used as the support for the matrix or gel, add a wad of glass wool to the dry column and secure it into position at the bottom of the column with a glass rod or piece of glass tubing. Add just above the glass wool one millimeter of clean fine grid silica sand. Next add buffer to the column (1/2 full) and check the apparatus for leaks. Close the exit valve of the column.

3. Pour an appropriate volume of gel suspension in order to fill the column to the required bed height (approximately 10 cm^3). This is done gently and slowly with the aid of a stirring rod having one end pressed gently against the inner wall of the column. The gel suspension is allowed to stream down the rod into the column. This procedure eliminates unnecessary turbulence and the introduction of air. Place any excess slurry in an appropriately labeled container.

4. Connect a buffer reservoir to the column and fill the column and reservoir with buffer. Open the outlet valve of the column and allow the buffer to flow through the column for approximately 20 minutes. Collect the buffer in a beaker. The flow of the buffer will assist the column in packing. Be certain to keep an adequate supply of buffer in the column so that the column never runs dry.

   The matrix or gel is packed when it stops compressing. The column should have a packed volume of approximately 10 to 12 ml of slurry. Close the outlet valve.

   This is a good place to stop. Cover the buffer reservoir and column outlet valve with paraffin and leave it until the next laboratory period.

Method, continued

Determine the Column Void Volume

5. Adjust the column flow rate by lowering or raising the buffer reservoir until the number of drops per minute is sufficient to permit 8 ml per minute of elution solvent to be collected.

6. Close the outlet valve and remove most of the buffer from above the column bed by using a pasteur pipet. Open the outlet valve and allow the buffer to penetrate into the gel or matrix. Be careful not to let the matrix become dry. Close the outlet valve.

7. Using a pasteur pipet, apply a protein solution whose molecular weight is known to be greater than the void volume of the matrix (Sephadex G-25 in this case) to the top of the column. The concentration of the applied sample should be about 0.5 mg/ml in a buffer volume that is equal to 1% of the total bed volume of the column. The volume of the column is equal to $\pi r^2 h$. One percent of the volume is obtained by multiplying the packed volume of the matrix by 0.01.

8. Open the outlet valve and allow the sample solution to penetrate the gel bed. Wash the remaining sample from the column wall by applying small amounts of buffer from a pasteur pipet while the flow of buffer from the column is carefully checked.

9. After the sample is applied, carefully pipet a 1 cm layer of buffer onto the top of the gel. Then completely fill the column with buffer and connect the buffer reservoir. Allow the elution to occur at the flow rate adjusted earlier.

10. Collect the column fractions in 1 ml volumes by using a test tube containing 1 ml of water as a reference. One should collect approximately 100 fractions. Analyze each fraction for the presence of protein using the biuret reagent. Compute the void volume by multiplying the volume collected in each fraction by the number of the fraction having the maximal absorbance for biuret complex with protein at 540 nm. The void volume is about one third of the total bed volume.

Dissolving the Sample

11. Dissolve the protein sample in the buffer used in the gel filtration column. The sample volume should be about 5% of the column bed volume. Repeat steps 5 to 10, above. If one is desalting a protein mixture, the volume of the sample can be as high as 25% of the total bed volume of the column.

Method, continued

The sample is applied in the same fashion as the blue dextran was applied.

12. Elute the sample with the gel filtration buffer. If one is desalting a protein solution, use either distilled water, ammonium hydroxide (0.03M) or acetic acid (0.05M) to elute the protein. Please Note: Proteins may precipitate in distilled water.

13. Recover the desalted protein by lyophilization to remove the low concentrations of acetic acid or ammonium hyforxide.

Results

The computed void volume of the column is:

____

The bed volume of the column is:

____

The number of milligrams of protein desalted is:

____

The percent purification:

____